Looking for Life, Searching the Solar System

How did life begin on Earth? Is it confined to our planet? Will humans one day be able to travel long distances in space in search of other life forms? Written by three experts in the space arena, *Looking for Life, Searching the Solar System* aims to answer these and other intriguing questions. Beginning with what we understand of life on Earth, it describes the latest ideas about the chemical basis of life as we know it, and how these ideas are influencing strategies to search for life elsewhere. It considers the ability of life, from microbes to humans, to survive in space, on the surface of other planets, and to be transported from one planet to another. It looks at the latest plans for missions to search for life in the Solar System, how these plans are being influenced by new technologies, and current thinking about life on Earth. This fascinating and broad-ranging book is for anyone with an interest in the search for life beyond our planet.

PAUL CLANCY is a former strategic planning manager within the Human Spaceflight Directorate at the European Space Agency in Paris. His recent duties included planning for exobiological research and the human exploration of Mars, and for European utilisation of the International Space Station.

ANDRÉ BRACK is Director of Research Emeritus in the Centre of Molecular Biophysics at the Centre National de la Recherche Scientifique, Orleans, France. His research focuses on the chemistry of the origins of life, and the search for life in the Solar System. He co-founded the European Exo/Astrobiology Network Association in 2001.

GERDA HORNECK is a former Head of the Radiation Biology Section and Deputy Director of the Institute of Aerospace Medicine at the German Aerospace Centre DLR in Cologne, Germany. She has coordinated several national and European projects in space biology, biological UV-dosimetry, biosensorics and biomarkers, and has coordinated human survivability in space studies.

Looking for Life, Searching the Solar System

Paul Clancy, André Brack
and Gerda Horneck

CAMBRIDGE UNIVERSITY PRESS
Cambridge, New York, Melbourne, Madrid, Cape Town, Singapore,
São Paulo, Delhi, Dubai, Tokyo

Cambridge University Press
The Edinburgh Building, Cambridge CB2 8RU, UK

Published in the United States of America by Cambridge University Press, New York

www.cambridge.org
Information on this title: www.cambridge.org/9780521124546

© P. Clancy, A. Brack & G. Horneck 2005

This publication is in copyright. Subject to statutory exception
and to the provisions of relevant collective licensing agreements,
no reproduction of any part may take place without the written
permission of Cambridge University Press.

First published 2005
This digitally printed version 2009

A catalogue record for this publication is available from the British Library

ISBN 978-0-521-82450-7 Hardback
ISBN 978-0-521-12454-6 Paperback

Additional resources for this publication at www.cambridge.org/9780521124546

Cambridge University Press has no responsibility for the persistence or
accuracy of URLs for external or third-party internet websites referred to in
this publication, and does not guarantee that any content on such websites is,
or will remain, accurate or appropriate.

R0427918448

Dedicated to the memory of David Wynn-Williams

Contents

Preface

This book has three intertwined threads which have inexorably come together at this, the dawn of the twenty-first century. Never before has the question of the origin and the destiny of life been stronger in the public and the scientific mind. Although the question 'What is life and how did it start on Earth?' has proven difficult to answer, despite significant research over recent decades, at least we now know a lot more about its molecular machinery than we did when Harold Urey and Stanley Miller did their famous amino acid generating experiments 50 years ago. And this should help us greatly in looking for the signatures of life in an extraterrestrial context. The discovery of such signatures might eventually allow us to answer the question 'How did life originate and where did it come from?' And if there is extraterrestrial life, can we find it out there in the Solar System?

The second thread is then the issue of whether or not humans are capable of long-duration interplanetary flights and long-term survival on inhospitable planetary surfaces.

The third thread sets the question: if humans *are* capable of such voyages, do we have a real, active role to play in the search for life in far-flung planetary environments within the Solar System?

The authors share a passionate belief that the destiny of humanity is strongly linked to its outward exploration of the vast spaces and resources of the Solar System and ultimately the Galaxy. They also share a strong belief that the public at large is fascinated by this subject, and is broadly aware that it could have enormous consequences for the future of our children and grandchildren. The authors also believe that the public is keen to know how terrestrial life fits into the vast astronomical picture now painted by modern science, how life originated and how widespread it is in the Universe. The authors have played leading roles in recent studies carried out in a European context in key areas such as exobiology and the search for life in the Solar System, as well as the survival prospects for humans in long-duration spaceflight and on planetary surfaces. They feel that the results of these studies merit exposure to a wide audience.

This book is not about an evangelical call to aggressively colonise and exploit other planets of the Solar System, but rather to explore them in a spirit of respect, and recognition of the central place played by diversity in the rich phenomenon of life. It covers a wide range of subjects across the spectrum of issues related to the exploration of and search for life in the Solar System. As a result, the material covered has a wide range of accessibility for the reader.

Readers interested in the general challenges and motivations for exploration will find Parts I, IV and V the most relevant to their interests. Those readers looking for the latest information regarding our present knowledge about the chemical basis of life, and how that relates to the survival of life in hostile environments as might be found on other planets, as well as the universal signatures of life, will find all this in Chapters 2, 3 and 5 in Part II. Chapter 4 covers the latest ideas about the possible traffic of living organisms between planets.

Part III deals with what we now know about how humans can survive in space, and potentially on the surfaces of other planets, and how humans could be directly involved in the search for life on other planets as well as the ethics involved in those searches. Parts IV and V round out the story by explaining the technologies that will be needed for this exploration, how the exploratory programme itself could play out in the coming decades and how the philosophical, societal and economic impacts of this exploration will be felt.

Acknowledgments

The authors would like to express their thanks to a wide variety of people in the European exobiology community who in one way or another, by dint of joint work, conversation, question or simply raised eyebrow, contributed to the dialogue that resulted in this book. They are: Gerhard Kminek, Oliver Angerer, Didier Schmitt, Marc Heppener, Jorge Vago, Dietrich Venneman, Bruno Gardini, Marcello Coradini, Peter Schiller, Franco Ongaro, all at the ESA, as well as Richard Bonneville, Sylvie Leon and Francis Procard at CNES. Special mention must also be made of the fine and indispensible contributions through the 'Red Book' – Exobiology in the Solar System and the Search for Life on Mars, ESA SP-1231 – that forms part of the intellectual backbone to this book. Here the contributions of Brian Fitton, Patrick Forterre, Colin Pillinger, Manfred Schidlowski, Heinrich Waenke, Francois Raulin, Beda Hofmann, Gero Kurat, Gian Ori, Nicolas Thomas, Frances Westall, Daniel Prieur and David Wynn-Williams were invaluable.

The other part of the backbone of this book is the insights into the potential survival of humans in long-term spaceflight and in carrying out planetary exploration. This was investigated as part of the HUMEX study carried out for ESA by a German Aerospace Centre (DLR) team and reported in ESA SP-1264 and special mention must be made of the contributions of Rainer Facius, Michael Reichert, Petra Rettberg, Wolfgang Seboldt, G. Reitz, C. Baumstark-Khan and R. Gerzer, all of DLR; Bernard Comet and Alain Maillet of MEDES in Toulouse; H. Preiss and L. Schauer of EADS (Germany); Dieter Manzey of the Technical University of Berlin; C. G. Dussap and L. Poughon of the Université Blaise Pascal-Clermont (France) and A. Belyavin of QinetiQ (United Kingdom).

Part I
The imperative of exploration

From the perspective of determinism and constrained contingency that pervades the history of life... life and mind emerge not as the result of freakish accidents, but as natural manifestations of matter, written into the fabric of the Universe.

Christian de Duve (1995)

1 Exploration as metaphor

The urge to explore has been ingrained in human nature since the emergence of our species. The story of the vast movements of peoples across oceans and continents is woven into the history of humankind. For millennia this drive has been at the core of European culture and history. The Greeks and Romans of antiquity explored distant regions of Asia and Africa far from their established empires. Vikings from Scandinavia and, quite probably, Irish monks reached the shores of North America centuries before Christopher Columbus set out on his voyage of discovery across the 'Ocean Sea'. In the fourteenth century Marco Polo penetrated the Chinese empire. In the centuries that followed, Henry the Navigator sponsored the vast Portuguese exploration of the coast of Africa, Cartier explored Canada and Magellan made the first perilous circumnavigation of the globe. The eighteenth and nineteenth centuries saw the great voyages of discovery of Captain Cook in the Pacific and the exploration of large swathes of Central and South America by Humboldt, who described the regions' plants and climate and studied the Earth's magnetic field. Charles Darwin's famous voyage around the world aboard the *Beagle* provided the evidence which would lead him to the theory of evolution. Nearer our own times, polar explorers have numbered such European pioneers as Nansen, Shackleton, Amundsen and Scott.

In the twentieth century, the mantle of pioneering exploration fell to the United States and the Soviet Union in their superpower rivalry, opening the way to outer space when the tiny Sputnik 1 forged its way across the world's skies in October 1957. Four years later, Yuri Gagarin became the first human in space with his single-orbit flight around the Earth, demonstrating that where robots would first go humans would soon follow. Unmanned and, later, manned missions to our Moon followed, and by the end of the century Pioneer 10 had photographed and flown past all of the great and outer planets and left the Solar System for the emptiness and darkness of deep interplanetary space. We seem now to stand at the threshold of a great leap forward in the story of human exploration and new questions crowd forward:

- What does our present knowledge about the nature and origins of life tell us about the existence and pervasiveness of life in the Solar System?
- What are the limits to human survivability in interplanetary space and on planetary surfaces?
- Can we identify and prepare the development of key technologies to enable such human Solar System exploration?
- Can we expect to find other non-terrestrial life forms in the Solar System and, if so, how should we look for them and would they pose a threat to human exploration?
- Can we devise an ethical planetary exploration approach within existing international accords that reflects civilised values?
- What exactly is the unique role of humans in the exploration of the Solar System in pursuit of the search for life?

Some of the great voyages of discovery might be able to give us clues about how to approach these questions.

Magellan

In the early part of the sixteenth century, the intense competition between Spain and Portugal to discover and exploit the lands to the west and south of Christopher Columbus's 'Ocean Sea' came forcefully to the attention of Pope Alexander the Sixth. The Pope drew a line on the map of the Atlantic and ceded the eastern portion to Portugal and the western part to Spain. Ferdinand Magellan, a Portuguese mariner who could not get the support of the Portuguese king, ended up sailing under the flag of the King of Spain after convincing the royal court at Seville that a Spanish expedition sailing west could find a way through the islands into the ocean beyond.

FIGURE 1.1 Ferdinand Magellan. (Courtesy of Mariners Museum, *Newport News.*)

He set sail from Seville on 10 August 1519. Of the five ships that left Seville that day only one ship, the *Vittoria*, returned to Seville after the first circumnavigation of the globe, on 6 September 1522. Magellan himself had been killed in the Philippines trying to convert the natives to Christianity and had discovered and named the Pacific Ocean. Within a span of just over three years, Magellan's voyage had

FIGURE 1.2 Magellan discovers the Straits of Magellan. (Courtesy of the Mary Evans Picture Library.)

determined the size and overall composition of our planet in terms of its two major oceans and the manner by which our globe can be circumnavigated. No man before or after, with the possible exception of Cook, has had a comparable effect on the exploration of the Earth, and settled forever the questions of its size and overall geography.

Captain Cook

On Friday 12 July 1776, Captain James Cook set sail from Plymouth in search of the North West Passage. He had renamed the three-masted collier under his command, *Resolution*, a fitting tribute to his state of mind in setting out on such a long and potentially perilous journey. It would involve stops in Tenerife, Table Bay and Queen Charlottes Sound in New Zealand before even the first mile of exploration could be embarked upon. The other ship in his command was also renamed by Cook. It took the name *Discovery*, and would join *Resolution* later on. No clearer names could have been chosen to portray Cook's determination to discover the hypothetical northern passage between the Pacific and the Atlantic Oceans. Cook had been engaged in voyages of exploration since 1772. Two- and three-year voyages of discovery were not unusual for explorers of the eighteenth century. Similar journey times there and back should not be daunting for human explorers of Mars in the twenty-first century.

The exploration of foreign places requires human qualities which transcend the values generally held in esteem at the conclusion of the twentieth century, a century which valued cleverness over fortitude, insight over intuition, calculated over hunch-based risk-taking, corporate over individual enterprise. The explorers of the twenty-first century will need to rediscover some of the values which drove men to take the personal and corporate risks which they did in the seventeenth and eighteenth centuries. Whereas Captain Cook's voyages appeared on the face of them to be neutral voyages of discovery, there was always the British Admiralty behind their financing. They were never simple voyages of discovery, but rather voyages of discovery in the furtherance of the interests of English trade and enterprise. So there is a lesson to be learned here.

Successful voyages of discovery have a back-story of exploration and discovery which will, if successful, mightily benefit those who have invested in them. Cook, the British Admiralty and others believed that the discovery of the North West Passage would facilitate trade between England, China, the Indies and the Pacific Islands and that English trade and commerce with the rest of the world would be greatly enhanced.

So we arrive at an interesting question: are there general lessons to be learnt from the voyages of exploration of the seventeenth and eighteenth centuries which could be applied to the space voyages of exploration of the Solar System in the twenty-first century? We will focus on:

- What were/are the arguments both for and against the enterprise? In Parts IV and V we will examine some of the arguments put for and against expanding human exploration out into the Solar System.
- What would be the logistical and psychological similarities? Hammond Innes, the writer of a convincing fictional account of Cook's last voyage, imagined Cook writing of his stay at the isolated Kerguelen Islands as follows:

  'The activity of both ships was so great that day that I doubt whether any man really believed it to be Christmas Day. And the next day was no better, the filling of the casks with water having to be completed and all to be rowed back and forth between beach and ship, also grass for the cattle to be conveyed in the same manner after being cut from small sheltered areas at the end of the inlet where it did grow. All this work, and the killing of the seals and birds, was much hampered by rain, which was so heavy that the bare rock of the hillside was a sheet of water and every gully a raging torrent.'

Can we imagine similar logistical and psychological stresses in human voyages of discovery in the Solar System? Cook was clearly expert at using as far as possible those resources he found along the way. In Part IV we will look at present thinking about so-called '*in-situ* resource utilisation' in planetary exploration. In Part III we will examine the physiological and psychological stresses on the human system in long-duration exploration missions.

- Concerning the nature of exploration itself, Innes imagines Cook writing: 'I have always referred to the ease of handling of these ships. There is none built anywhere that I have found so entirely suited to the purpose of surveying uncharted and dangerous coasts. This I have proved, never at any time wanting a better. Slowness of sailing is no disadvantage in this work. What is important is to be able to stop the ship dead and turn her in her own length, which they do very readily. I am told they look clumsy, but that is only to those that have not sailed in them or who do not understand what is required of the work on which I have been engaged these last 8 years.'

  Can we imagine similar statements from future astronauts about the qualities of their ships? What is the importance of technology in exploration. Does it play a decisive role, as clearly it did in Cook's case? In Parts IV and V we will deal with this.

An epitaph to Cook was made by his surgeon onboard the ship, Samwell, as follows:

'With clear judgement, strong masculine sense and most determined resolution – with a genius peculiarly turned for enterprise, he pursued his objective with unshaken perseverance; – vigilant and active in eminent degree; – cool and intrepid among dangers; patient and firm under difficulties and stress; fertile in expedients; great and original in all his designs; active and resolved in carrying them into execution.'

We ask, can our future explorers of the Solar System have these qualities?

It is necessary in all this to reflect on what the past can tell us about the future. The voyages of Magellan and Cook were both marked by searches for passages to other realms, the Straits of Magellan in the former case and the North West Passage in the latter.

Voyages of discovery can never be completely planned. Magellan and Cook, although they knew in general terms what they wanted to explore and discover, were never completely sure what they would find and how that would change and inform their future activities and decisions. Future human space exploration will have to proceed in a way which allows the enterprise and initiative of private individuals and companies as much freedom as possible to invest in space exploration and reap the benefits of those investments. The days of the monopolistic domination of all aspects of space exploration by space agencies, the manipulation of space activities for national propaganda purposes, and the use of billions of dollars and euros of taxpayers' money to further the ambitions of politicians will have to give way to a time when space becomes a real resource for humanity in general.

Charles Darwin

It was 27 December 1831. A cloudy and calm day broke over Plymouth. During the morning the wind freshened and at 2 p.m. Captain Robert FitzRoy came aboard his 30-metre vessel, HMS *Beagle*, together with the 22-year-old naturalist, Charles Darwin, ready to depart from Devonport for a voyage of survey. The voyage eventually took about five years and brought the *Beagle* and its crew around the globe; in the first year, from the British Isles across the Atlantic to South America, where they spent more than 3 years on excursions to various sites, then around Cape Horn and up the Pacific coast of South America to

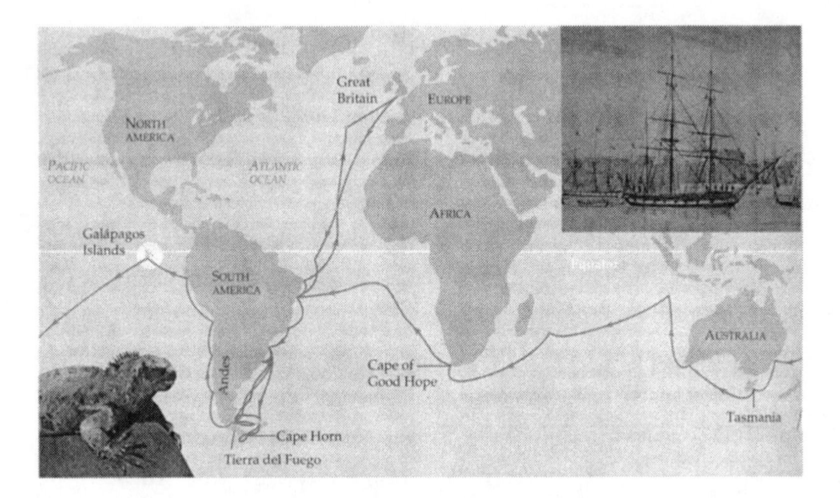

FIGURE 1.3 HMS *Beagle* in the Straits of Magellan and *Beagle* voyage (taken from *Darwin and the Beagle*, Penguin Books, 1969).

the Galapagos Islands. After crossing the Pacific to New Zealand, they visited Tasmania and Australia, crossed the Indian Ocean and passed into the Atlantic by way of the Cape of Good Hope, and returned to England at Falmouth, where they arrived on 2 October 1836. They made a wealth of new and exciting discoveries about the exotic plants and animals they found, with the most stunning at the Galapagos Islands. Yet nobody in 1836 realised that the voyage of the *Beagle* and the discoveries Darwin had made, especially on the Galapagos Islands, would lay the foundations for the formulation of the theory of evolution. Moorehead in his account of Darwin and the *Beagle* has written:

> *But now here on the Galapagos, faced with the existence of different forms of mocking birds, tortoises and finches on different islands, different forms of the same species, he was forced to question the most fundamental contemporary theories ... Darwin's thesis was simply this: the world as we know it was not just 'created' in a single instant of time; it had evolved from something infinitely primitive and it was changing still.*

Charles Darwin's voyage on board the *Beagle*, and his discoveries during the voyage which finally led to the fundamental theory of evolution, is one of the most illuminating examples of how exploration serves as a pathfinder for shaping human cultural and scientific perspectives. In the endeavour to explore the Earth, humans have crossed the seas, climbed the highest mountains, visited the poles and studied the deepest parts of the oceans. Now, humans have conquered the furthest limits of habitability on our planet. If this urge to explore is ingrained in human nature and if it will impel future human generations to continue on this path, what will result as we set out on the first steps of human exploration of the Solar System?

Amundsen and Scott

We are in January 1912, just past midsummer again at the Antarctic. The Sun is off its maximum height as it slowly starts to head for the horizon and the descent into winter in late February. Five exhausted Britons struggle across the ice pulling their sledges behind them. They are expectant and slightly hurried, despite their exhaustion. Their leader urges them on – they are near their goal now. It is early afternoon. Up ahead they see an object – a man-made object. As they draw closer it is clear that it is a tent. The leader's heart begins to sink. Gradually things become clearer. The small tent has a pole emerging from its peak and attached limply to the pole is a flag. A flag with a red background and an offset blue and white cross. A Norwegian flag. Robert Falcon Scott and his British expedition to be the first to the South Pole have been beaten to that goal by Roald Amundsen and his Norwegian team, who had placed the Norwegian flag at the South Pole on 14 December 1911. In semi-secrecy, Amundsen had launched and carried out a highly successful and logistically efficient assault on one of the last bastions of the wilderness on Earth. Whereas Scott had disdained the use of dogs for his expedition, Amundsen started with 97 dogs and used them in a highly efficient and, according to some people, 'mechanistic' manner as a resource for both transportation and food. Basically, when the dogs became exhausted with pulling the sledges carrying the men and supplies, they were slaughtered and eaten. The British considered this to be ungentlemanly and unsporting. Yet Amundsen made it first to the South Pole and returned with his expedition intact. Scott was beaten to the Pole and afterwards he and all his men perished in a blizzard within 15 kilometres of a cache of food, supplies and shelter. There can be no clearer conclusion from this that, in the realm of exploration, logistics are paramount in determining the chances of success or failure.

Robert Zubrin, who we will see in this book has proposed a striking 'Mars Direct' approach to human exploration of Mars using *in-situ* resources, has waxed eloquent on the merits of Amundsen's philosophy. He has drawn conclusions about how this could be generalised to the wider context of exploration of the Solar System. Amundsen's 'live off the land' approach which had served him so well in his Arctic expedition of 1903–1905 appeals to Zubrin as a model or metaphor of how humans should approach the exploration of Mars. In 1905, Amundsen had been the first to traverse the North West Passage and his survival owed much to his ability to live like the local Inuit people and eat the freshly killed meat of caribou, which guaranteed life for the expedition members. This 'live off the land' philosophy as an efficient and inherently safe way of conducting expeditions in harsh and exteme conditions greatly impressed Amundsen and he used it in his successful South Pole expedition where the men killed and ate their dogs. Proponents of direct Mars missions like Zubrin argue strongly for a so-called '*in-situ* resource utilisation', and in particular the conversion of carbon dioxide in the Martian atmosphere to methane to provide fuel for the return journey to Earth.

Ernest Shackleton

On Friday, 1 August 1914, with war clouds darkening over Europe, Ernest Shackleton and a team of 56 left London in the barquentine-rigged *Endurance*, headed for the Antarctic. Shackleton planned the first trans-Antarctic expedition. In order to find a way through the icefields to Vahsel Bay on the Antarctic landmass, from where he intended to start the trans-Antarctic traverse, Shackleton sailed eastward from South Georgia to the South Sandwich Islands and then southward into the Weddel Sea. But against all expectations the ice was dense that year and by 15 January, just off Vahsel Bay, *Endurance* was firmly frozen into the pack ice. Throughout the Antarctic winter, *Endurance* and her crew drifted westwards and northwards, away from the 77 degree south latitude which had been her most southerly achievement. Blizzards constantly pounded *Endurance*, dumping up to 100 tons of snow on her decks and up against her exposed sides. The rumbling of shifting and rafting ice ridges that bore down on the ship was a constant reminder of the powerful and unremitting forces of Nature in whose jaws they were trapped. Food for the men and dogs became scarce as the local supply of seals and penguins dried up. The ice continuously threatened them. Shackleton wrote, 'The noise resembles the roar of heavy, distant surf. Standing on the stirring ice one can imagine it is disturbed by the breathing and tossing of a mighty giant below.'

The *Endurance* drifted for over 1000 miles in these conditions, but as temperatures rose the ice became more active and started to crush the ship. On 23 October the ship was breached and began to leak. Four days later they abandoned her and set up camp on the ice nearby. A month later *Endurance* sank beneath the ice. With the ice now melting and crumbling beneath them, Shackleton and his men and dogs headed towards Paulet Island nearly 350 miles away. They hauled three of *Endurance*'s boats with them, ready to cram the 28 men into when the ice disintegrated. Moving from cracking-up ice floe to ice floe they drifted past the Antarctic Circle on New Year's Eve, a year after crossing almost the same latitude on their outward journey. Alternating between stretches in the boats when the water opened a little and back on ice floes when it closed again they reached Elephant Island in the South Shetland Islands on 12 April. It was their first encounter with land in 16 months. But it was deserted land and Shackleton decided to send a party on the dangerous crossing to South Georgia, 800 miles away, to seek help. After great difficulty in launching the *James Caird* open boat, Shackleton and five others set

FIGURE 1.4 The *Endurance* trapped in the ice. (Courtesy Scott Polar Research Institute.)

out across the storm-swept southern ocean on Easter Monday, 1916. Navigating through the icebergs, the suffering men aboard the *James Caird* made good progress in the rough weather. After some days they were hit by an Antarctic storm, with freezing sea spray covering the boat with ice which they had to hack away in case it sink. They started to suffer frostbite. At midnight on the eleventh day they were struck by a huge cresting wave. After they had recovered from the chaos they realised that the boat was half-full of water and threatened to sink, so a furious bout of bailing out was needed. At midday on 8 May they saw the black cliffs of South Georgia and huge waves crashing over the rocks at the base of the precipitous cliffs. Suffering greatly from thirst, and with night coming on, Shackleton and his crew could see no place to make a safe landing. They would have to wait until next morning. During the night they were struck by hurricane force winds and drifted back out to sea. At dawn no land was in sight and they had to recover lost ground as the winds died. But again, when they approached, no safe landing site could be seen. They had to spend another gruelling night at sea. But the next day the winds died completely and after a fierce battle with the angry reefs Shackleton and his men made it into King Haakon Bay.

But they were on the wrong side of the island. The Stromness whaling station was nearly 20 miles away, over nearly 5000-foot high mountains. The men being very weak, Shackleton left three of them with the boat where they had landed and with the remaining two set off over the glacier-covered mountains. After an overnight struggle to cross the mountains and against the overwhelming desire to sleep, which would have meant certain death, they reached the whaling station at midday, to the amazement of the whalers stationed there. Shackleton, with the aid of a whaling vessel, picked up the three men on the other side of the island the next day. In the succeeding months, he attempted, first from South Georgia and then with vessels out of the Falklands and Punta Arenas, to rescue his men stranded on Elephant Island, but was repeatedly beaten back by the pack ice and by engine failure. Finally, on the fourth attempt using the Chilean steamer *Yelcho*, he reached the stranded men, finding them all safe and well. They had endured 105 days on the lonely island.

What can be learned from Shackleton's extraordinary expedition? Principally, it must be the importance of the ability of the commander to provide outstanding leadership as well as make effective decisions, the safety and wellbeing of the crew being the uppermost concern. Shackleton's remarkable endurance and his firm leadership during all phases of the expedition, right through to the rescue of the men stranded on Elephant Island, is without question one of the greatest feats in the annals of exploration. Also, his ability to make effective decisions far from the eyes of any 'mission control' was the primary reason for the survival of all his men. From the moment they left South Georgia in December 1914, to the moment Shackleton and his two men stumbled into the Stromness whaling station in May 1916, the expedition might well have been on another planet. The commanders of human exploration missions to other planets of our Solar System may well have to display similar feats of endurance and effective leadership, far away from the ability of any Earth-based Mission Control to influence events.

Given the brilliant and daring exploits of these explorers of the past, we are tempted to ask: has exploration reached a dead end now that we have seen the limits of our planet? Or will new frontiers be opened beyond the surface of Earth? After having reached the most distant and hostile places on Earth, is it not a logical next step to start exploring the neighbourhood of the Earth, the Moon and the planets of the Solar System?

The era of human spaceflight started more than 40 years ago. On 12 April 1961, the Vostok 1 spacecraft carrying Yuri Gagarin orbited the Earth. But the environment in Earth orbit is hostile, and these excursions carry many life-threatening risks. In order to survive in space, humans need to carry along their own life support systems, either provided by the spacecraft or – during extravehicular activity – by a spacesuit. Even with these support systems humans cannot escape all the adverse conditions of spaceflight. Cosmic radiation and weightlessness impair human health and efficiency in space. Given these hazards, we ask what induces in certain people the urge to leave the comfortable regimes of the home planet and submit themselves to the hostile conditions of outer space?

At the beginning of the era of human spaceflight in the early 1960s the answer to this question was simple: politics. At that time human space activities were driven by the competition between the two large political blocks: the USA and the Soviet Union, both vying for 'firsts' in human spaceflight. The race culminated on 20 July 1969, when the Apollo 11 mission placed humans on the Moon. In the course of its lifetime the Apollo programme placed 12 humans on the lunar surface and has been highlighted by R. Johnstone as *'a major national event'*.

However, political rivalries are not the only driving force for the pursuit of human spaceflight. Gerda Horneck and others have described how other drivers rooted in science, technology, culture and economics have inspired us to explore low Earth orbit, the Moon and Mars using robotic probes.

From a science point of view, the Moon offers a rich potential as a scientific outpost for science *of the Moon*, which includes geophysical, geochemical and geological research, leading to a better understanding of the origins and evolution of the Earth-Moon system. Science *from the Moon* takes advantage of the stable lunar surface, its atmosphere-free sky and radio-quiet environment for astronomical observations, especially on the far side of the Moon. Science *on the Moon* provides opportunities to study the stability of biological and organic systems under hostile conditions, especially radiation, the regulation of autonomous ecosystems, and the scientific and biomedical preparation of human missions to other planets of the Solar System, as has been described by H. Balsiger and others.

Mars is a major target in the search for life beyond the Earth and the only planet – except the Earth – located within the habitable zone of our Solar System. Dry river beds indicate that huge amounts of water and a denser atmosphere were present about 3 billion years ago. During this warmer and wetter period, life may have originated on Mars and may even exist today in special 'oases or ecological refuges', such as geological formations below the surface that have favourable conditions for life. As has been pointed out by CNRS researchers André Brack and Frances Westall, the recent controversial claims of the discovery of possible fossil life forms in Martian meteorites mean that the search for morphological or chemical signatures of life or its relics is one of the primary and most exciting goals in the exploration of Mars.

In addition to exobiology, research in geology, mineralogy and atmospherics will play a central role in the scientific exploration of Mars. The general goal is to understand planetary formation and evolutionary processes including, if possible, the evolution of life. Mars is the planet most similar to the Earth, and the question of climatic change, especially the loss of water and atmospheric gases, may be linked to the evolution of the Martian climate. Studies in this area may also contribute to our understanding of the history and future of the terrestrial climate, as well as to the understanding of the evolution of the Solar System as a whole. Human intervention may well be crucial in achieving many scientific goals, such as site identification by local analysis, sample acquisition at these sites, sampling and – when a laboratory is available on Mars – sample analysis, directly on the Martian surface.

New and demanding challenges usually require new technologies. Transporting humans away from their home planet to establish new and habitable environments on the Moon or on Mars will be just such a new and demanding challenge.

But how will we test these new technologies? The Moon will be an excellent test bed for technologies required for future space exploration, such as missions to Mars. The environments on the Moon and Mars are similar, above all, the radiation field, thin atmosphere, reduced gravity and the presence of dust. Therefore human missions to the Moon will be extremely useful in preparation for journeys to Mars. In addition, the Moon may provide resources for *in-situ* exploitation, such as oxygen for life support and fuel production, or ^{3}He as a fuel for fusion reactors on Earth, producing very low radioactivity by-products.

Meeting the scientific objectives of Mars missions will require autonomous and intelligent tools, such as intelligent sample selection and collection systems, with a high degree of automation and robotics. As soon as human crew are involved, the need for integrated advanced sensing systems will arise, such as bio-diagnostics, medical treatment, and environmental monitoring and control. New technologies developed in these areas will almost certainly have wide applications on Earth for more prosaic purposes. Furthermore, the need for technologies for *in-situ* resource utilisation and, above all, for producing propellant from atmospheric CO_2 or water ice – and also for life support purposes – will be a powerful driver for the development and testing of advanced technologies.

The Moon, as companion of the Earth, has shaped the conditions on the Earth, helping to make it a habitable planet with a sustainable biosphere. It is a natural space station orbiting the Earth. Human voyages to the Moon foreseen in Jules Verne's *From the Earth to the Moon* captured imaginations in both the nineteenth and twentieth centuries. Establishing a first habitable outpost beyond the Earth on a natural body of our Solar System will be an important cultural event. It will re-shape the human perspective away from an Earth-oriented one and towards a more Universe-oriented one. Lunar bases first set up for scientific and technological purposes will develop into tourist attractions. Once relatively economic access to space becomes available, space tourism will become a major business.

Mars has occupied a similar place in the human imagination. When the first telescopes were directed towards Mars, channels and shaded areas were interpreted as huge agricultural plantations or lichens covering the surface. The supposed canals were seen as desperate attempts to bring water to desertifying equatorial regions. With the crater-filled images sent back by Mariner 9 and the hostile and chemically highly reactive surface encountered by the two Viking spacecraft that landed on Mars in 1976, these ideas were quashed. But Viking and the follow-on missions also told us that the early Mars had probably experienced a similar climate to that of the early Earth, when life started here. This has prompted NASA scientist Chris McKay and others to propose Mars as a suitable and attractive target for terraforming, by using, for example, modern techniques of planetary and genetic engineering.

Human spaceflight has the potential of promoting extensive peaceful cooperation on a global scale. The International Space Station (ISS) is the largest international science and technology project ever undertaken. This cooperative venture for the joint development, operation and utilisation of a permanent space habitat in low Earth orbit involves a significant proportion of the spacefaring nations. With the ISS, a new era of peaceful cooperation in space on a global scale has started. The major partners are the USA, Russia, Japan, Europe and Canada, with the USA taking the leading role. The lessons learned from this experience could be crucial for future large international space projects such as missions to Mars.

From an economics point of view, human exploratory missions with the Moon or Mars as targets are relatively expensive. M. Reichert and others have estimated the cumulative costs of a human Moon

FIGURE 1.5 The International Space Station. (Courtesy ESA.)

programme, lasting for about 20 years, at about $50 billion, with a mean annual budget of about $2.5 billion. For an exploratory programme comprising three human missions to Mars over 20 years, nearly the same cumulative costs have been estimated, namely $52 billion with an average budget per year of $2.6 billion. On the other hand, a comparable annual budget is presently spent by the USA just for the operation of the Space Shuttle fleet. Hence, a lunar base or a Mars programme are relatively affordable, especially if carried out with costs and tasks shared internationally. Synergies with terrestrial applications are expected in various fields, such as robotics and nanotechnology, as well as autonomous health control systems and/or telemedicine systems. However, even considering the possibility of *in-situ* resource exploitation and the prospects of a future boom in lunar tourism, it will require a major investment which probably will have to be shared by many nations.

Long-duration missions beyond low Earth orbit clearly present tremendous human challenges. Recent studies on the survivability and adaptation of humans to long-duration interplanetary and planetary environments, such as the HUMEX study (*Hum*ans on *Ex*ploratory Missions), have revealed the human health and survival issues involved. The study has been reported by G. Horneck and others. With the emphasis on human health care, wellbeing and performance, the study has looked at the effects of radiation and zero-gravity, as well as psychology issues. Health maintenance and the developments needed for advanced life support were also taken into account. This is dealt with in detail in Part III.

In Table 1 are given some of the motivations and consequences of explorations from the past to the present day. On the motivations side, a wide range of objectives can be at play, such as desire for fame and riches, scientific curiosity (although this latter is usually less than the scientific community vociferously declares). The desire for conquest, political or even military objectives may also be involved. On the results side, there has been a similarly wide range of consequences, ranging across geographical discovery, the creation of wealth and technological

TABLE 1 Some motivations and results of exploration.

Explorer	Motivation	Results
Columbus	• Finding a short way to the Indies for increased trade • Search for gold	• Unintended discovery of America • Questionable colonisation of the West Indies • Failure to follow through to mainland America
Magellan	• Exploitation of navigable ways around Earth • Conversion to Christianity of heathens found on way	• Large-scale geographical definition of Earth • Discovery of Straits of Magellan
Captain Cook	• Voyages of discovery and acquisition • Exploration of Pacific • Search for North West Passage	• Detailed definition of Pacific geography
Beagle/Darwin	• Voyage to discover plant and animal species in Pacific and Atlantic	• Discovery, cataloguing of 100s of new species • Theory of Evolution
Amundsen	• To be first to South Pole	• First to South Pole, 1912

LEO Space Missions 1961–1969	• Cold War rivalry to demonstrate conquest of space • To put first men on Moon and return them to Earth	• Proof of survival of humans in space • Proof of EVA
Apollo Space Missions 1969–1973	• To carry out science survey • To prove US technological superiority over USSR	• Survival of men on Moon • Return of Moon rocks to Earth • Confirmation craters result of meteor impacts
Planetary Missions 1976–2003	• Robotically explore Mars, Giant and Outer Planets • Substitute for more expensive human missions	• No life on surface of Mars • Discovery of volcanoes on Io • Discovery of Jupiter's rings • Discovery of complexity of Saturn's rings and resonance of moons • Unexpected diversity of Solar System
LEO Human Spaceflight 1973–2003	• Zero-gravity research • Continuation of human spaceflight for strategic reasons (USA) • Development of space stations MIR, Salyut (Russia)	• Zero-gravity phenomena • Better understanding of hazards to astronauts due to zero-gravity and radiation

progress, the establishment of empires and the spread of philosophies and religions. Negative consequences, such as the decimation of native populations and the encouragement of rivalrous wars have also, sadly, in some cases been part of the outcome.

We move now towards certain conclusions regarding those attributes that successful and progressive exploration programmes must have.

Firstly, the mission/exploration programme must have clearly laid out objectives known and understood by all key players, both at the home base and at the front line. This is a strategic goal. The success of the Apollo programme derived in large measure from the absolute clarity of its goal, as so eloquently declared by President Kennedy.

Secondly, the logistics and logistical support must be optimised and must be consciously worked out, planned for and implemented in a manner capable of supporting, in a clear and convincing way, the primary objectives laid out in the first point above.

Thirdly, the tactical pursuit of the mission/exploration programme must be flexible enough to allow on-the-spot explorers and managers to make changes to the plan in order to pursue newly emerging scientific/technological goals as the exploration proceeds, intelligently and *within* the constraints laid down in the logistical plan under the second point above and the overall objective, under the first point above.

Fourthly, as far as possible, all planning should involve a high degree of efficiency based on what might be called the Amundsen/Zubrin 'live off the land' approach, using *in-situ* resources to produce a wide range of consumables such as energy, fuel, water and, possibly, food.

Lastly, the programme must have a strong ethical context derived from internationally agreed ground rules derived from treaties such as the United Nations Treaties on Outer Space. This ethical context must have as its primary motivations the following: firstly, the protection of other planets and bodies from danger and damage, whether physical, chemical or biological, and the protection of the Earth – especially from potential pathogens coming from outside.

Secondly, the respect for human life and the reflection of that in the protection of spacefaring humans, as far as possible, from dangers such as radiation, skeletal degeneration due to zero gravity and potentially hazardous non-terrestrial pathogens. Thirdly, a respect for all non-terrestrial organisms and species discovered in the course of the exploration and, in particular, the application of measures to ensure that human activities do not threaten the ecologies of such species, or drive them to extinction.

CONCLUSION

Since the emergence of modern man in Africa a few million years ago, the history of our species has been a story of mass migrations and the colonisation of increasingly greater tracts of the Earth's surface, driven by the human urges for exploration and settlement. Whether the urge to explore is ingrained in our genes or not, as claimed in some of the more colourful arguments for a furious pursuit of space exploration, is a rather moot point given our present inability to trace any particular behaviour of our species to specific features of our genetic make up.

Thus, it may be difficult or even impossible to identify exploratory behaviour as a survival factor in any Darwinian sense. Yet the rewards of exploration as a key feature of human activity and wired into our history, mythology and the story of human progress seem evident. In this sense, exploration is perceived collectively by humanity as a fundamental activity bearing on the ability of our species to survive and thrive in an essentially uncertain and potentially hostile Universe.

The early history of humankind provides ample evidence of this. Humans crossed the ice bridge of the Bering Straits from Asia into America and settled the two American continents 50 000 years ago. More recently, in the last thousands of years, once-nomadic tribes swept westward from Central Asia and settled eastern, central and western parts of Europe, ensuring that the majority of the apparently diverse European languages have a common Indo-European heritage.

In more recent times, Viking marauders raided northern and central parts of Europe but ended in the peaceful settlement of wide regions of France, Italy and the Danube basin. Around the same time, Mongol, Tartar and barbarian hordes surged westward into Europe from Asia, precipitating the fall of the declining Roman Empire. Asian peoples pressed eastwards from East Asia and from the Japanese archipelago into the great island universe of Oceania in the Pacific.

By the fifteenth and early sixteenth centuries, the spirit of the Renaissance in Europe, and all its political, social, economic and even philosophical ramifications had worked their way into the hearts and minds of ambitious men, pushing them to pursue dangerous exploratory forays into parts of the world which had previously been unknown to Europeans. The triple motif of 'God, Gold and Glory' gripped the imaginations of men, propelling many, such as Columbus, Magellan, Da Gama and the Spanish and Portuguese explorers of America, to take huge risks in pursuit of these ends.

This composite of motivations had as its most salient feature the prospect of enhanced trade. This was the easiest to sell to potential sponsors, such as emperors, kings, princes and potentates, who held the purse strings in Renaissance Europe and who understood the connection between the vigorous pursuit of trade and its impact on the wealth of their own state and, by implication, their own power, prosperity and influence in the world.

By this time Europe had developed a singular appetite for luxury goods, such as spices to fortify and preserve foods, and rich fabrics to adorn the persons of wealthy merchants and their families. Silk from Asia and spices such as cinnamon, cloves and nutmeg from the East were in high demand. But the sea routes were dominated by Arab middlemen who could exact enormous bounties and prices for these goods from their European clients. Eventually these clients felt impelled to find other routes, such as a westward way, to the Spice Islands. In parallel, an emerging mercantile class created a huge demand for transactions based on money rather than barter and, in consequence, the need for precious metals such as silver and gold as the physical embodiment of this monetary form of exchange.

Exploration and the risks it entails require extraordinary commitment and leadership as the explorers face into and take on the unknown. A shared philosophy and world outlook, either stated or unstated, is almost mandatory. For the Spanish and Portuguese explorers of the fifteenth and sixteenth centuries it was the religion of Catholic Europe. This extraordinary Christian religion of redemption and immortality was forced on the hapless so-called savages encountered on the way by these explorers and missionaries. In this case, exploration and philosophy were inseparable.

Without a strongly held philosophical view of the world, those involved in exploration can rarely achieve the cohesion and unity of purpose needed to face down the enormous physical, psychological and financial challenges they will face on a daily basis. Although we abhor the excesses and genocidal tendencies of the Spanish conquistadors, we cannot deny the power of a shared view of the world as part of the psychological reality of exploration allowing small groups of only a few hundred men to literally take on whole countries. In the particular circumstances of the Catholic invasion of South America the mania of missionary zeal led to a thirst for fame, glory and even sainthood. This crucial desire for image, glory and prestige are rarely far from the core of exploratory activity. The US/USSR race to the Moon in the 1960s is a particularly stark example of this.

With the coming of the Enlightenment in the late sixteenth and seventeenth centuries the 'God, Gold, Glory' motif metamorphosed into a less rampant trinity derived from rationalist enlightenment philosophy. The benefits of trade and the fame to be seized through philosophical and scientific achievement now took root in the minds of men. God, gold and glory gave way to philosophy, trade and science. This latter trinity of goals still prevails today.

The Enlightenment above all brought science, and the belief in the primacy of data and its rational interpretation. The pursuit of systematic knowledge based on theory, and its proof or disproof by experimental and empirical methods, became the chief mechanism for establishing truth. Much of this relied on physical measurements made by newly developed and/or perfected instruments such as the magnetic compass, the sextant and especially the development of accurate chronometres, crucial to the accurate measurement of longitude, as exemplified by John Harrison's maritime clocks. This, in turn, led to the creation of the world's first maps accurate to a few tens of metres. The flowering of science in the Enlightenment quickly found its way into a whole range of human activities, spurring the inception of the industrial revolution but above all the development of faster and better-performing ships which were crucial to the further pursuit of human exploration.

The urge to explore is an impulsive one and not a cerebral one. It is probably rooted very deeply in our human neurological make up. It is constantly pushing against the boundaries set by the constraints of urban life and security. Perhaps our exploratory urges are connected with survival not only of individuals but survival of the species as a whole. If that is the case, we will need to pay attention that turning away from commitments to planetary exploration does not lead to longer-term dangers for the human race. As M. E. DeBakey has eloquently put it:

> *We know from earliest recorded history, some 5000 years ago, that human beings have always sought to learn more about their world. The century just ended has witnessed stunning advancements in science and medicine, including the launching of space exploration. There is a danger, however, that the new century may usher in an age of timidity, in which fear of risks and the obsession with cost-benefit analysis will dull the spirit of creativity and the sense of adventure from which new knowledge springs.*

Part II
How can we know life and its origin?

Anyone who tells you that he or she knows how life started on the sere earth some 3.45 billion years ago is a fool or a knave. Nobody knows. Indeed, we may never recover the actual historical sequence of molecular events that led to the first self-reproducing, evolving molecular systems to flower forth more than 3 million millennia ago. But if the historical pathway should forever remain hidden, we can still develop bodies of theory and experiment to show how life might realistically have crystallized, rooted, then covered our globe. Yet the caveat: nobody knows.

Stuart Kauffmann (1995)

Introduction

Stuart Kauffmann is a member of the Santa Fe Institute and a leading thinker on self-organisation and complexity in the field of biology. His caveat above is fundamental. It serves to reinforce and underline one of the purposes of this book. Although we will look at the fundamental chemistry and biochemistry of life, we will not be doing so to lay out a firm and refutable theory of the origins of life. Rather, we want to identify those crucial aspects of life's chemistry and biochemistry which can help us to understand the linkage between that chemistry and that of the planets of the Solar System, and of visitors from outside the Solar System – comets. We will lay out a roadmap on how to conduct intelligent strategies for the search for life on other planets. What we are concerned with is identifying those crucial and fundamental chemical and biochemical features of life that allow us to recognise its traces – living or dead – in other parts of the Solar System. That this may give hints as to life's origins we consider a bonus.

A priori, we would imagine that researchers working in the fields of origins of life, artificial life and extraterrestrial life should know what they are searching for and/or trying to reproduce in their laboratories. This is presently not the case, since no formal definition of life has up to now been commonly accepted by the scientists working in these fields. This applies even for a definition of *a minimal life*, the simplest possible form of life as has been described by P. L. Luisi. A recent survey carried out by the International Society for the Study of the Origins of Life asking for a definition of life produced 78 answers, from which it was clear that there is no definition that satisfies the entire community. These results have been described by G. Palyi. Perhaps the most general working definition is that adopted in October 1992 by the Exobiology Programme within NASA: 'Life is a self-sustained chemical system capable of undergoing Darwinian evolution.' Implicit in this definition is the fact that the system is driven by external matter/energy provided by the environment. Modern viruses are not autonomous (they depend upon more sophisticated living organisms for their reproduction) and do not, therefore, fulfil the NASA definition. Amazingly enough, no consensus can be reached even now as to whether these are to be considered as living or not. For the sake of simplicity, we will not contribute any further to the semantic complexity by adding another definition of life. We will just consider two characteristics needed by matter to be considered alive: self-reproduction and evolution.

2 The molecular basis of life on Earth

In this triumphant era of molecular biology and the first draft of the human genome, one might have supposed that we would know the answer to the question, What is life? Yet we do not. We know bits and pieces of molecular machinery, patches of metabolic circuitry, genetic network circuitry, means of membrane biosynthesis, but what makes a free-living cell alive escapes us. The core remains mysterious.

Stuart Kauffmann (2002)

At the present time, terrestrial life is the only reference we have. From what we know of such life and from the point of view of chemistry, life can be defined as an open chemical system able to (i) make more of itself by itself (self-reproduction) and (ii) to evolve. Self-reproduction is, from the point of view of chemistry, a form of autocatalysis. In this sense, a set of selected molecules is self-amplified via self-selection and becomes a *peculiarity* in its environment. Life, as such a chemical peculiarity, modifies its environment and generates anomalies in the Earth's atmosphere, minerals and oceans. Therefore, any existing or even extinct life could be inferred to exist if it produces peculiarities on a large scale which are distinguishable from the environment in which the life exists or existed in the past. For example, the composition by elements of the Earth's biological material (the biomass) is totally different from that of the Earth itself, and from that of the cosmos as a whole, as shown in Table 2 below.

All present-day living systems on Earth use cells as self-reproducing building blocks. Again, from the point of view of chemistry, there are three features of these building blocks which appear to be paramount:

TABLE 2 Relative abundances of the biogenic chemical elements CHON (adapted from A. Brack, 1990).

	Relative abundances (in mass %)		
Elements	**Biomass**	**Earth**	**Cosmos**
Carbon	17.9	0.005	0.34
Hydrogen	10.5	0.003	70.0
Oxygen	71.0	29.9	0.92
Nitrogen	0.23	<0.0001	0.12
Helium	<0.0001	<0.0001	28.0
Other elements	0.37	70.09	0.62

- The cell, in order to self-reproduce and to allow evolution, must contain a store of information. This is the genetic message that is stored in nucleic acids in the nucleus.
- The basic chemical work is carried out by protein enzymes.
- The self-reproductive machinery is isolated from the surrounding environment by a chemical barrier.

Taking the first of these, we know that nucleic acids are long chains built from building blocks called nucleotides. Each nucleotide is composed of a base, a sugar and a phosphate group.

The sugar is either a ribose or a deoxyribose. When the chain is built from nucleotides with two OH groups in positions $2'$ and $3'$ on the sugar (a ribose), as shown in Figure 2.1, then this is a ribonucleic or RNA chain. When the chain is built from nucleotides with the OH group missing from the $2'$ position on the sugar (a deoxyribose), we have deoxyribonucleic acid or DNA.

In Figure 2.2 is shown a nucleotide in which the base contains two cyclic elements. This is called a purine base and there are two of them involved in RNA and DNA, namely adenine A and guanine G. When the base contains only a single cycle these are called pyrimidines and again there are two of them involved with RNA and DNA,

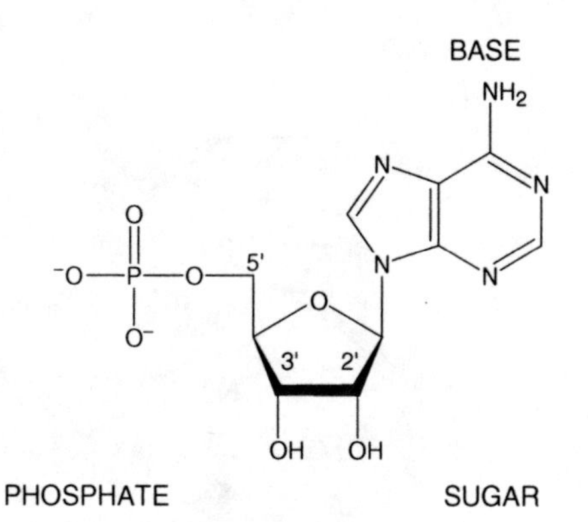

FIGURE 2.1 A typical nucleotide.

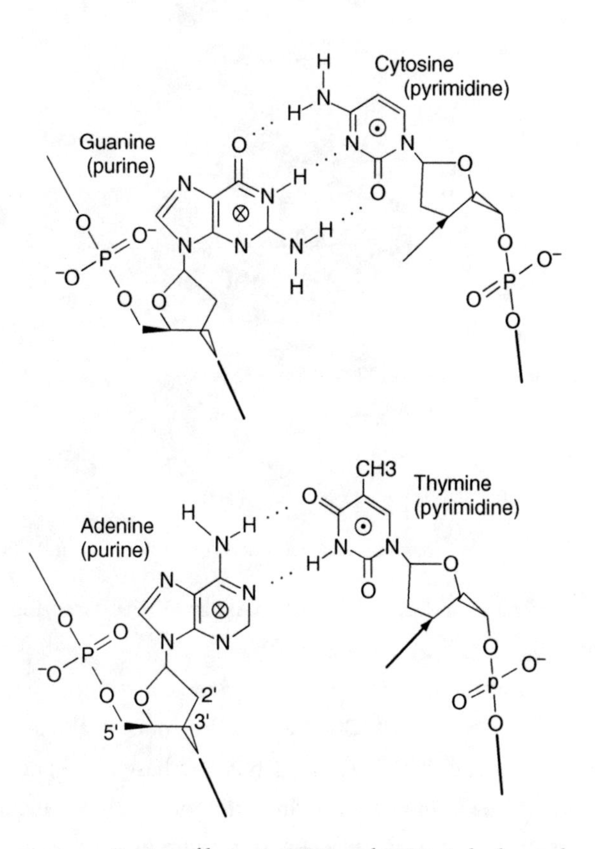

FIGURE 2.2 Pairing of bases in DNA and RNA via hydrogen bonds.

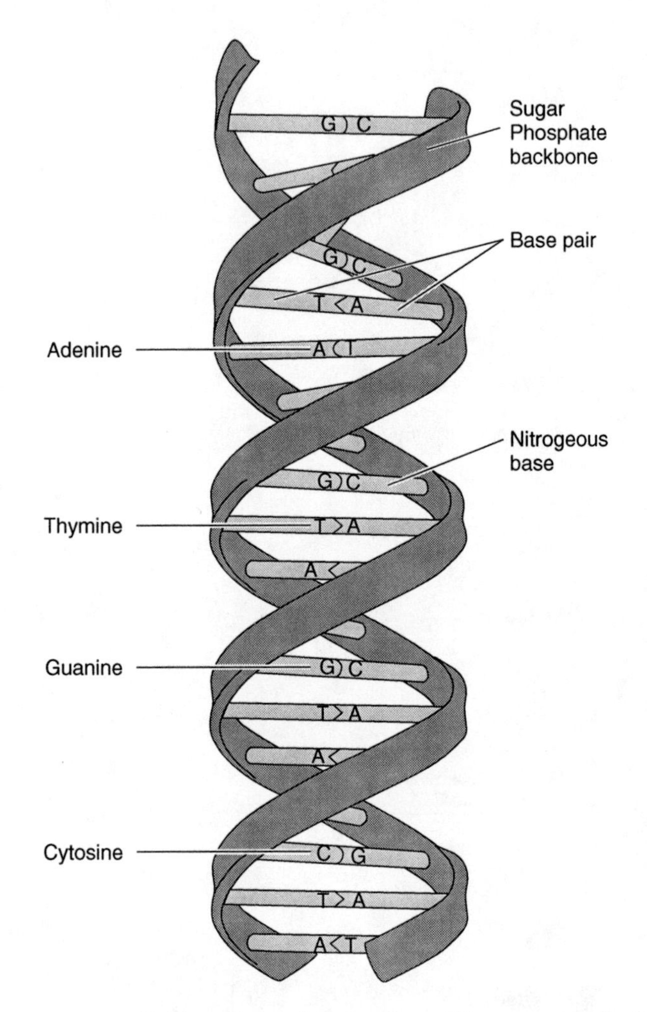

FIGURE 2.3 Pairing of bases C to G and A to T between strands of DNA.

namely cytosine C and uracil U in RNAs. In DNAs, uracil is replaced by thymine T.

The cyclic ribose in the nucleotides possesses four asymmetric carbons in positions 1′, 2′, 3′ and 4′. This makes the nucleotides globally asymmetric. In DNA and RNA the bases can pair up (base pairing) via hydrogen bonds. A hydrogen bond is a weak electrostatic chemical bond that forms between the hydrogen atoms of the OH, NH, SH groups and atoms of oxygen, nitrogen or sulphur.

Base pairing of A to T and G to C in DNAs, and A to U and G to C in RNAs is the key recognition process in the biological self-replication of nucleic acids and drives the organisation of nucleic acids in asymmetrical double helices. This means that in DNA, adenine normally pairs with thymine and guanine normally pairs with cytosine.

The DNA and RNA are therefore right-handed double helices that are bound together by the pairing of bases between the helices right along their length. The self-replication comes when the helices unwind from one another. We could imagine this happening by cutting from left to right, say, through the chain shown in Figure 2.3 between the paired bases. When the two helices are separated, the guanine on the left of the top strand attracts a cytosine, the next cytosine attracts a guanine, the next adenine attracts a thymine and so on. Therefore each single strand creates a new partner strand that is a copy of the original strand, so the original DNA has produced a new copy of itself.

How is all this important? Since proteins are the building blocks of life and since there are over a million different proteins involved in life, a mechanism is needed for the construction of individual proteins. Proteins are built up from amino acids. There are 20 different amino acids of the form NH_2-CHR-COOH used by life to construct proteins. They differ from each other by the nature of the side-chain -R, which is either a single hydrocarbon chain or a hydrocarbon chain ended by a chemically reactive function like -NH_2 or -COOH.

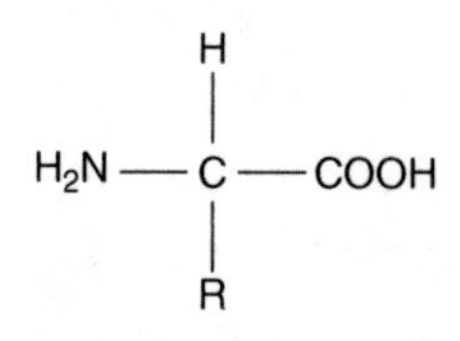

Now each amino acid has a signature. It is a word consisting of three letters, and the letters are taken from the DNA alphabet, namely the G, C, A, T letters. For example, the sequence GCA codes for the amino acid alanine. So when GCA appears on the DNA, the cellular

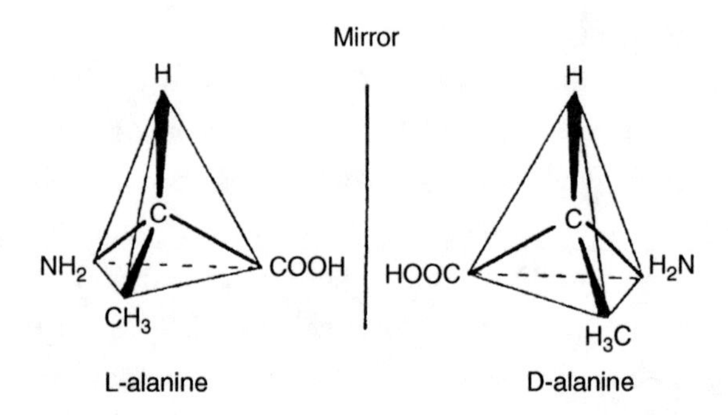

FIGURE 2.4 Left- and right-handed mirror image versions of the amino acid alanine.

manufacturing process knows that alanine is required at that point, to be used in whatever protein is being manufactured.

The central carbon atom of an amino acid occupies the centre of a tetrahedron and when the four constituents at the four summits are different, the amino acid is one-handed and exists in two forms, a left-handed form, noted L, and a right-handed form, noted D. Each form is a mirror image of the other. The example of the amino acid alanine is shown in Figure 2.4.

The 20 natural protein amino acids, with the exception of glycine which is not asymmetric, belong to the L-series. This means that all proteins on Earth use only left-handed amino acids to build the protein building blocks. Therefore, life on Earth is one-handed, and this is referred to as 'homochirality', from the Greek '*kheiros*' or hand.

When amino acids join together to form proteins they do so through peptide bonds, where the nitrogen atom in the NH_2 group on one amino acid links up with the carbon atom of the COOH group of another amino acid (see Figure 2.4). This means that the protein backbone is a repeat of N-C-C atoms like ... N-C-C-N-C-C-N ... etc.

But the N-C-C angle is about 110 degrees, so the atoms do not line up in a straight line. Instead, they form either sheet-like structures or helical structures, as shown in Figures 2.5 and 2.6.

The third main feature we have noted is that the genetic message needs to be isolated from its surrounding environment. This can

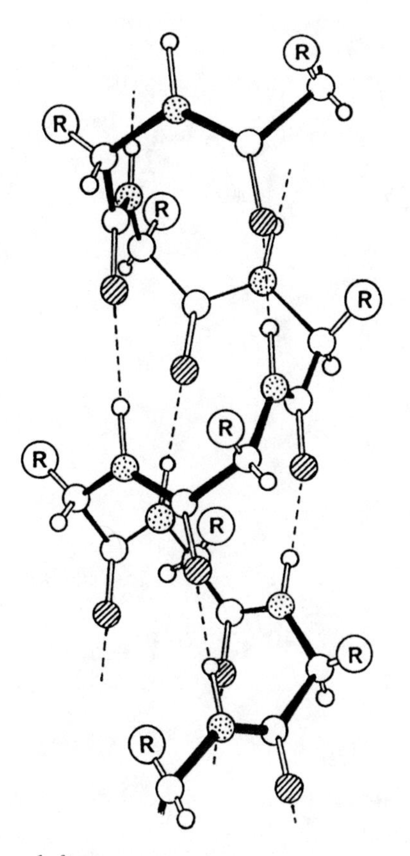

FIGURE 2.5 α-helices.

be achieved by a biological membrane. Research has shown that biological membranes are made of molecules that have a hydrophobic (i.e. water-repelling) hydrocarbon tail and a hydrophilic (i.e. water-attracting) polar head (see Figure 2.7).

These molecules can form spherical micelles, bilayers or liposomes when placed in water, since the tails will repel water and seek to join each other, whereas the heads will seek to interface with the water (see Figure 2.8).

All these structures are very interesting, in the sense that they offer a confined environment protected from the surrounding water in which the genetic material could be stored.

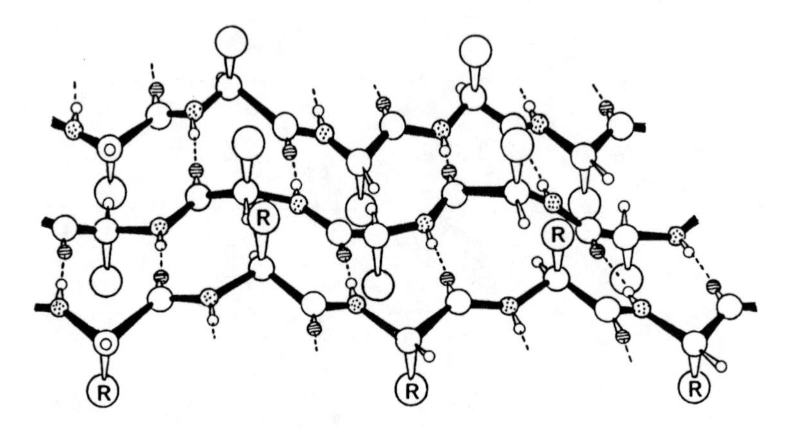

FIGURE 2.6 β-sheets.

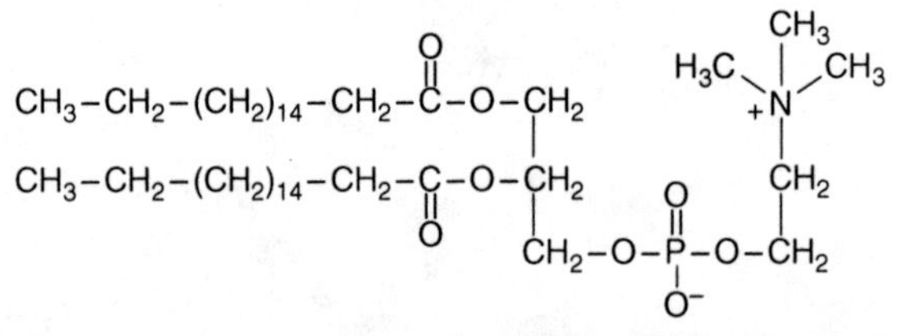

FIGURE 2.7 Phosphatidylcholine, a naturally occurring phospholipid found in biological membranes. A phospholipid is composed of two hydrocarbon chains attached to a phosphate group via glycerol, CH_2OH-$CHOH$-CH_2OH.

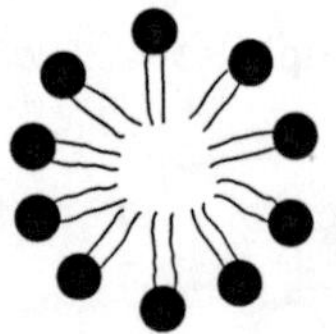

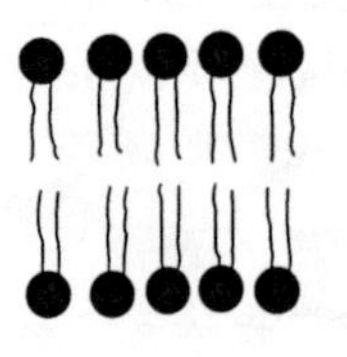

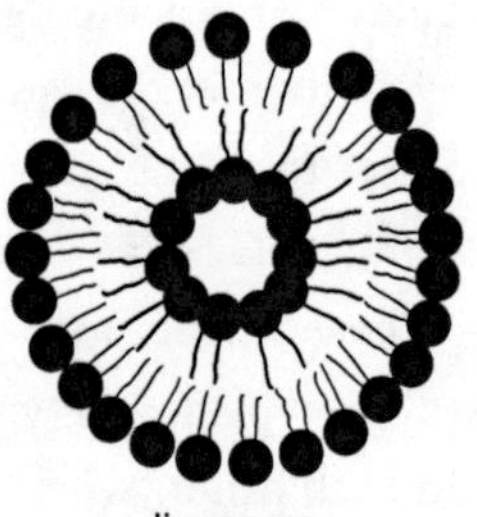

FIGURE 2.8 Micelles etc.

We now turn to the chemical specific features of a carbon-based life. As we have seen, the constituents of life are self-reproducing (or autocatalytic in chemistry terms) and are also able to evolve. In order to do this, the molecules bearing the hereditary memory must be capable of being extended in space, and diversified in structure and function by appropriate chemical reactions. This can best be achieved with a scaffolding of polyvalent atoms like pieces of Lego. In the light of what is now known, carbon chemistry is, by far, the most productive in this respect.

Interestingly, radio astronomy has now detected about 90 carbon-containing molecules in the interstellar medium, while only 9 silicon-based molecules have been detected. So it would seem that the generalisation that carbon-based life is universal is not just an anthropocentric view.

Biological carbon exhibits two distinctive features relative to non-biological carbon: one-handedness and stable isotope distribution. As we have seen, in an amino acid the carbon atom occupies the centre of a tetrahedron. When the four constituents at the four summits are different, the amino acid exists in two possible forms which are mirror images of each other, a left-handed form and a right-handed form. Theoretical studies have shown that autocatalytic systems fed with both left- and right-handed molecules must become one-handed in order to survive. Therefore, one-handedness of complex organised molecular systems or homochirality represents probably the only universal basic signature of life.

The uptake of carbon dioxide by living systems can produce biological molecules enriched in ^{12}C carbon isotopes at the expense of ^{13}C isotopes. For example, on Earth, over 1600 samples of fossil kerogen (a complex organic molecule produced from the debris of biological matter) taken from sedimentary rocks have been compared with carbonates in the same sedimentary rocks. The organic matter is enriched in ^{12}C by about 2.5 %. This difference is sometimes taken to be one of the most powerful indications that life on Earth was active nearly 3.9 billion years ago because the samples include specimens from across the whole of the Earth's geological timescale, but this is still much debated.

Specific features of carbon chemistry also exist at the molecular scale. For instance, the reaction:

$$n\,CO + (2n+1)\,H_2 \rightarrow C_nH_{2n+2} + n\,H_2O$$

produces a large set of hydrocarbons when carried out in a non-biological context.

However, the hydrocarbons present in biological membranes contain components which have non-random carbon number distributions, such as straight-chain hydrocarbons, with a pronounced carbon number maximum which is different from that obtained by the abiotic reaction above, or where the range of carbon numbers is restricted, or where there is an even over odd or odd over even carbon number predominance.

Concerning the anomalies introduced by life into the environment, the presence of oxygen in the atmosphere of a planet could indicate the presence of life at the surface. Methane can also be an indicator of life as it could signify the existence of subterranean hydrogen-fed micro-organisms called methanogens which, on Earth, are known to produce methane biologically. Micro-organisms also leave both large-scale (macroscopic) and small-scale (microscopic) mineral signatures, either as fossilised mineral deposits or as biologically produced minerals from microbial build-ups, as we will see in Chapter 5.

Liquid water is extremely important in relation to the need for mobility in biological systems. As elements of an open system, the constituents of a living system must be able to move or diffuse at a reasonable rate. This cannot happen in the solid state. In a gas, the fast diffusion of such constituents is possible, but only a limited number of volatile organic molecules are stable. A liquid phase, such as liquid water, offers the best environment for the diffusion and the exchange of dissolved organic molecules. Other solvents could be considered, such as liquid ammonia, hydrogen sulphide and sulphur oxide, as well as hydrocarbons, organic acids or alcohols. Compared to any of these possible solvents, however, liquid water exhibits many special properties, including a good stability in relation to heat and a relatively good resistance to UV radiation.

In the Solar System, liquid water is a fleeting substance that can persist only above 0 °C and under a pressure higher than 6 mbars. According to its molecular weight, water should be a gas under standard terrestrial conditions by comparison with compounds such as CO_2, SO_2, H_2S, etc. The liquid state of water exists due to the ability of the water molecule to form hydrogen bonds (see Figure 2.9). This is not restricted to water molecules since alcohols exhibit a similar behaviour. However, the way in which a network of water molecules can form via hydrogen bonds is so tight that the boiling point of water is raised from 40 °C (which is a temperature inferred from the boiling point of the smallest alcohols) to 100 °C.

The temperature range of liquid water can be enlarged by depressing the freezing point below 0 °C. For instance, salts dissolved in water (brines) depress the freezing point; the 5.5 % (by weight) salinity of the Dead Sea depresses the freezing point of sea water by about 3 °C. Large freezing point depressions are observed by using 15 % LiCl (23.4 °C) or by 22 % NaCl (19.2 °C). Monovalent and divalent salts are essential for terrestrial life because they are required as co-catalysts in many enzymatic activities associated with biological processes. Usually, the salt concentrations that can be tolerated are quite low

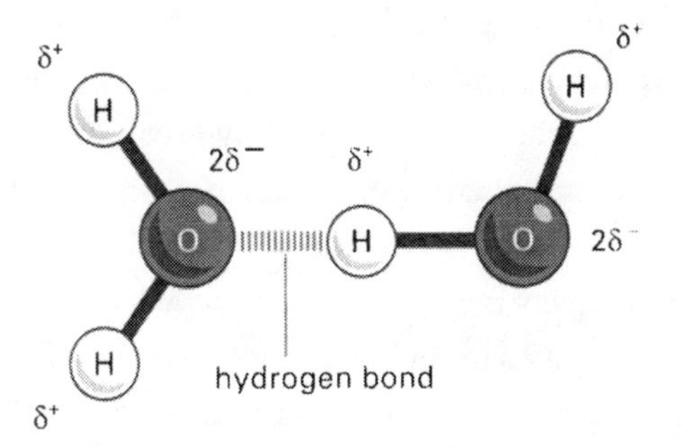

FIGURE 2.9 Typical formation of hydrogen bonds between hydrogen atoms and nearby oxygen, nitrogen or sulphur atoms.

(<0.5 %) because high salt concentrations disturb the networks of ionic interactions that shape long chain biological molecules (biopolymers) and hold them together. However, salt-loving micro-organisms, known as extreme halophiles, tolerate a wide range of salt concentrations (1–20 %) and some have managed to thrive in hypersaline biotopes (salines, salt lakes) of up to 25–30 % sodium chloride.

Liquid water has probably been permanently present at the surface of the Earth thanks to the size of our planet, its distance from the Sun and other factors. If the planet had happened to be much smaller, like Mercury or the Moon, its gravity would have been too weak to retain any atmosphere and, therefore, any liquid water. If the planet had been too close to the Sun, the mean temperature would have been too high due to the intensity of the sunlight. Any sea water present would have evaporated and delivered large amounts of water vapour to the atmosphere, thus contributing to the greenhouse effect. Such a positive feedback loop can lead to a runaway greenhouse effect whereby all of the surface water is transferred to the upper atmosphere. Here, photodissociation by ultraviolet light breaks the molecules into hydrogen, which then escapes into space, and oxygen, which recombines in the crust.

However, the Earth has retained its permanent liquid water thanks to its constant greenhouse atmosphere. Water could risk provoking its own disappearance. The atmospheric greenhouse gas CO_2 normally dissolves in the oceans, and is finally trapped as insoluble carbonates by rock-weathering. This negative feedback is expected to lower the surface pressure and the temperature to an extent that water would be largely frozen. However, on Earth, active plate tectonics and volcanism recycled the carbon dioxide by breaking down subducted carbonates. In the end, the size of the Earth, its distance from the Sun and the presence of plate tectonics meant that the planet never experienced either a runaway greenhouse effect or a divergent glaciation.

Water participates in the production of clays. Clay minerals are formed by the reaction of water with silicate minerals. As soon as liquid water was permanently present on the Earth's surface, clay minerals accumulated and became suspended in the water reservoir.

The importance of clay minerals in the origins of life was first suggested by J. D. Bernal. For Bernal, clays were interesting because of (i) their ordered arrangement, (ii) their large adsorption capacity, (iii) their shielding properties against sunlight, (iv) their ability to concentrate organic chemicals, and (v) their ability to serve as templates on which polymerisation could take place. Since the seminal hypothesis of Bernal, many prebiotic scenarios involving clays have been developed and many prebiotic experiments have been conducted using clays. They will be described in the chemical section.

To recapitulate the importance of water for life, it is clear that liquid water is intimately associated with all modern living systems. It is also commonly agreed that liquid water played a major role in the appearance and evolution of life. As a solvent, water favoured the diffusion and exchange of organic molecules. Liquid water has many additional pecularities. Water molecules establish hydrogen bonds with prebiotic molecules containing both hydrophilic and hydrophobic groups. In water, large organic molecules self-organise in response to their hydrophobic and hydrophilic propensities. This duality generates interesting prebiotic situations, such as the selective aggregation of mini-protein sequences of alternating hydrophobic – hydrophilic amino acids into thermostable geometries endowed with chemical activity. In addition to the H-bonding capability, water exhibits a large dipole moment as compared to alcohols, for example. This large dipole moment favours the dissociation of ionisable groups, such as $-NH_2$ and -COOH, leading to ionic groups which can form additional H-bonds with water molecules, thus improving their solubility. This is also true for metal ions, which have probably been associated with organic molecules since the beginning of life and generate interesting topologies. Liquid water was probably active in prebiotic chemistry as a clay producer and heat dissipator. Liquid water is also a powerful hydrolytic chemical agent. As such, it allows pathways that would have few chances to occur in another solvent.

Because of its specific set of properties, liquid water is generally considered as a prerequisite for terrestrial-type life to appear on a planet.

Saturn's moon Titan has a very active carbon chemistry but is too cold to host liquid water. It can therefore serve as a reference planetary laboratory to study, by default, the role of liquid water in exobiology.

It is therefore of primary importance to search for liquid water when searching for Earth-like life elsewhere. Other solvents, such as liquid ammonia, liquid hydrocarbons and alcohols, could be considered, but living systems metabolising in these non-aqueous solvents, if any, will necessarily be different from terrestrial life. Nevertheless, they cannot be discarded, a priori. They would probably be much more difficult to detect than extraterrestrial aqueous life, the identification of which already represents a tremendous challenge. (See Appendix 2 for more details on the properties of water.)

Turning now to the question of origins, we note that Charles Darwin first formulated the modern approach to the chemical origin of life. In February 1871, he wrote in a private letter to Hooker: 'If (and oh, what a big if) we could conceive in some warm little pond, with all sorts of ammonia and phosphoric salts, light, heat, electricity, etc., present that a protein compound was chemically formed, ready to undergo still more complex changes, at the present day such matter would be instantly devoured or adsorbed, which would not have been the case before living creatures were formed.'

For 50 years, the idea lay dormant.

In 1924, the young Russian biochemist Aleksander Oparin pointed out that life must have arisen in the process of the evolution of matter thanks to the reducing nature of the atmosphere, which would have been the most likely source of the reduced organic molecules that are involved today in biological processes.

In 1928, the British biologist J. B. S. Haldane, independently of Oparin, speculated on the early conditions suitable for the emergence of life. Subjecting a mixture of water, carbon dioxide and ammonia to UV light should produce a variety of organic substances, including sugars and some of the materials from which proteins are constructed. Before the emergence of life they must have accumulated in water to form a hot, dilute 'primordial soup'. Almost 20 years later, J. D. Bernal conjectured in 1949 that clay mineral surfaces were involved in the origin of life. In 1953, Stanley Miller obtained amino acids by subjecting a mixture of methane, ammonia and water to electric discharges. Miller's experiment opened the field of experimental prebiotic chemistry that aims to reconstruct primitive life in a test tube.

The question arises: can we consider reconstructing simple life in a test tube? As we have seen, it is difficult to define what is meant by the word 'life'. One generally considers as living any chemical system which is able to transfer its molecular information via self-reproduction and which is also able to evolve. The concept of evolution implies that the system normally transfers its information fairly faithfully but occasionally makes random errors, leading potentially to a higher efficiency and possibly to a better adaptation of itself to environmental changes. Schematically, the units of primitive life can be compared to parts of 'chemical robots'.

By chance, some parts self-assembled to generate robots capable of assembling other parts to form new robots identical to the mother robot. Sometimes, a minor error in the building generated more efficient robots, which then became the dominant species. By analogy with contemporary life, it is generally believed that the parts were

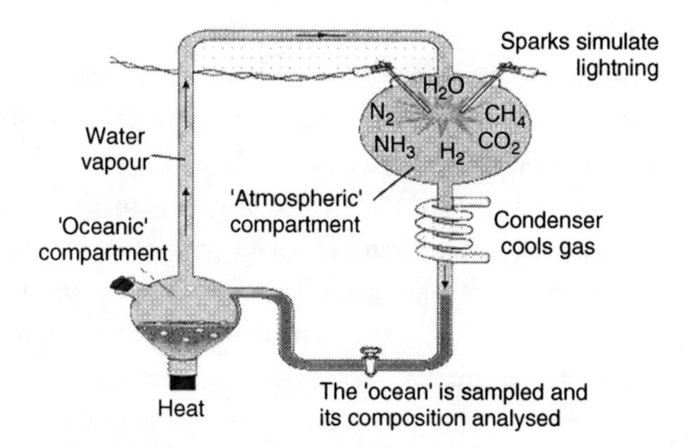

FIGURE 2.10 The Miller-Urey experiment (from Purves *et al.*, *Life: The Science of Biology*, 4th Edition, by Sinauer associates and W. H. Freeman).

made of organic matter, i.e. carbon skeletons flanked by H, O, N and S atoms, capable of mobility in liquid water.

The number of parts required for the first robots is still unknown. This number is expected to have been rather small because the chemical robots emerged when the Earth was under heavy bombardment by debris left over from the formation of the Solar System. A simple self-reproducing system would have been more robust, and offered better chances to resist the cataclysmic impacts during this heavy bombardment. If the number of parts was small, chemists have a reasonable chance to reproduce primitive chemical robots in a test tube. If the number of parts was very large, this mission of the chemists will be almost impossible.

As already mentioned, liquid water played a major role in the appearance and early evolution of life. Geological evidence suggests that the early Earth has been water-covered since perhaps as much as 4.4 billion years ago. This is on the basis of enriched ^{18}O isotope ratios measured in 4.4 billion-year-old zircons found in younger sediments from Western Australia. Zircons are very resistant crystals of zirconium silicates containing traces of uranium and thorium, which enable the dating of the rock. The chemical reconstruction of primitive life in test tubes must therefore be carried out in liquid water. It is also necessary to understand the possible sources of organic molecules involved in primitive life.

Originally, the carbon needed to construct the organic building blocks of life was available as simple volatile compounds, either reduced (methane CH_4) or oxidised (carbon monoxide CO or carbon dioxide CO_2).

As we have seen, the Russian biochemist Aleksander Oparin suggested that the small reduced organic molecules needed for primitive life were formed in a primitive atmosphere dominated by methane. In 1953, Stanley Miller identified four of the 20 naturally occurring amino acids which resulted after a mixture of methane, ammonia, hydrogen and water was exposed to electric discharges. Since this historic experiment, 17 natural amino acids have been obtained via the intermediate formation of simple precursors, such as hydrogen cyanide and formaldehyde. It has been shown that spark discharge synthesis of amino acids occurs efficiently when a reducing gas mixture containing significant amounts of hydrogen is used. However, even if the true composition of the primitive Earth's atmosphere remains unknown, geochemists now favour a non-reducing atmosphere dominated by carbon dioxide. Under such conditions, the production of amino acids appears to be very limited, thus contradicting Oparin's original hypothesis.

Deep-sea hydrothermal systems may also be potential environments for the synthesis of prebiotic organic molecules. Hydrocarbons containing 16 to 29 carbon atoms have been detected by N. G. Holm and J. L. Charlou in the Rainbow ultramafic hydrothermal system of the Mid-Atlantic Ridge. H. Yanagawa and K. Kobayashi found amino acids, although in low yields, under conditions simulating these hydrothermal vents. As we have seen, because of the high temperatures, hydrothermal vents are often discounted as efficient reactors for the synthesis of bio-organic molecules. However, the products that are synthesised in hot vents are rapidly quenched in the surrounding cold water, which may preserve the organics so formed.

If the carbon source for life was carbon dioxide, the energy source required to reduce the carbon dioxide might have been provided by the oxidative formation of pyrite from iron sulphide and hydrogen sulphide. Pyrite has positive surface charges and bonds the products of carbon dioxide reduction, giving rise to a two-dimensional reaction system, a 'surface metabolism' as reported by G. Wächtershauser in 1998. Laboratory work by C. Huber and G. Wächtershauser in 1997 and 1998 has provided support for this promising new hypothesis.

What about the possibility of an extraterrestrial source of organic molecules? Besides abundant hydrogen and helium, 114 interstellar and circumstellar gaseous molecules have presently been identified in the interstellar medium. It has also been conjectured that ultraviolet irradiation of dust grains may result in the formation of complex organic molecules. The interstellar dust particles are probably composed of silicate grains, surrounded by ices, including molecules which contain carbon.

In laboratory experiments by G. M. Munoz Caro and others, ices of H_2O, CO_2, CO, CH_3OH and NH_3 were deposited at 12 K and a pressure of 10^{-7}mbar and irradiated with an electromagnetic radiation representative of the interstellar medium. The solid layer that developed on the solid surface was analysed and, in order to exclude contamination, parallel experiments were performed with ^{13}C-containing substitutes. Sixteen amino acids were identified in this simulated ice mantle of interstellar dust particles. The results were confirmed by the ^{13}C experiments, which definitely excluded contamination. The chiral amino acids were found to be a 50/50 mixture of left- and right-handed forms. These results strongly suggest that amino acids are readily formed in interstellar space.

The incorporation of interstellar matter in meteorites and comets in the pre-solar nebula provides the basis of a cosmic dust connection. A comparison of interstellar and cometary ices using recent data from ESA's Infrared Space Observatory (ISO) has revealed important similarities between interstellar ices and volatiles measured in the coma of some comets.

Comets show substantial amounts of organic material, as was demonstrated by the ESA cometary mission Giotto to Halley's comet in 1986. As described by A. Delsemme, on average, dust particles ejected from the nucleus of Comet Halley contain 14 % organic carbon by mass. About 30 % of cometary grains are dominated by light elements C, H, O and N, and 35 % are close in composition to carbon-rich meteorites called carbonaceous chondrites. Among the molecules identified in comets are hydrogen cyanide and formaldehyde. The presence of purines, pyrimidines, and formaldehyde polymers has also been inferred from the fragments analysed by the Giotto Picca and Vega mission Puma mass spectrometres. However, there is at present no direct identification of the complex organic molecules present in the dust grains and in the cometary nucleus.

Many chemical species of interest for extraterrestrial life have been detected in Comet Hyakutake in 1996, including ammonia, methane, acetylene, acetonitrile and hydrogen isocyanide. In addition

FIGURE 2.11 ESA's Infrared Space Observatory (ISO). (Courtesy ESA.)

to the hydrogen cyanide and formaldehyde seen in several earlier comets, Comet Hale-Bopp was also shown to contain methane, acetylene, formic acid, acetonitrile, hydrogen isocyanide, isocyanic acid, cyanoacetylene and thioformaldehyde. Cometary grains might, therefore, have been an important source of organic molecules delivered to the primitive Earth. This has all been reviewed by P. Ehrenfreund and S. B. Charnley in 2000 and P. Ehrenfreund and others in 2002.

The study of meteorites, particularly the carbonaceous chondrites that contain up to 5 % by weight of organic matter, has allowed close examination of the extraterrestrial organic material that has been delivered to the Earth. Nucleic acid bases, purines and pyrimidines have been found in the Murchison meteorite by P. G. Stoks and A. W. Schwarz. G. Cooper and others found one sugar (dihydroxyacetone), sugar-alcohols (erythritol, ribitol) and sugar-acids (ribonic acid, gluconic acid) in the Murchison meteorite but no ribose, the sugar which links together the nucleic acid building blocks, as we have seen earlier in this chapter. Vesicle-forming fatty acids have been extracted from different carbonaceous meteorites by D. W. Deamer in 1985.

In 1998, J. R. Cronin and others reported finding eight amino acids which are used to build proteins in the Murchison meteorite, among more than 70 amino acids found in that meteorite. As we have seen, these amino acids are asymmetric. Usually, the two mirror-image configurations of the same amino acid are found in equal proportions. However, small (1.0–9.2 %) L excesses were found in six α-methyl-α-amino acids from the Murchison (2.8–9.2 %) and Murray (1.0–6.0 %) meteorites as reported by S. Pizzarello and J. R. Cronin. The presence of L excesses in these meteorites points towards an extraterrestrial process of asymmetric synthesis of amino acids, an asymmetry that is preserved inside the meteorite. These excesses may help to understand the emergence of a one-handed life.

Louis Pasteur was probably the first to realise that biological asymmetry (one-handedness) could well be the distinguishing mark between inanimate matter and life. There are strong biochemical and stereochemical reasons why life might favour this one-handedness or homochirality (see Appendix 3).

The excess of the one-handed amino acids, as found in meteorites, may result from the processing of the organic mantles of the interstellar grains from which the meteorite was originally formed. That processing could occur, for example, by the effects of circularly polarised synchrotron radiation from a neutron star, a remnant of a supernova. Strong infrared circular polarisation, resulting from dust scattering in reflection nebulae in the Orion OMC-1 star-formation region, has been observed by J. Bailey and others in 1998. In 2001, Bailey suggested that circular polarisation at shorter wavelengths might have been important in inducing this chiral asymmetry in interstellar organic molecules that could be subsequently delivered to the early Earth.

Dust collection in the Greenland and Antarctic ice sheets and its analysis show that the Earth captures interplanetary dust as micrometeorites at a rate of about 50–100 tons per day. About 99 % of this mass is carried by micrometeorites in the 50–500 μm size range. This value is much higher than the most reliable estimate of the meteorite capture, which is about 0.03 tons per day. In the case of Antarctic micrometeorites, a high percentage of unmelted chondritic micrometeorites from 50 to 100 μm in size have been observed, indicating that a large fraction entered the Earth's atmosphere without drastic heating or melting.

In this size range, the carbonaceous micrometeorites represent 80 % of the samples and contain 2 % of carbon, on average. This quantity of incoming micrometeorites might have brought to the Earth about 10^{17} tons of carbon over a period of 300 million years, corresponding to the late terrestrial bombardment phase as reported by M. Maurette and others in 1998 and 2000. For comparison, this delivery represents more carbon than that engaged in the surface biomass, i.e. about 10^{12} tons. These grains also contain a high proportion of metallic sulphides, oxides and clay minerals that belong to various classes of catalysts. In addition to the carbonaceous matter, micrometeorites might also have delivered a rich variety of catalysts, having perhaps acquired specific crystallographic properties during their synthesis in the microgravity environment of the early solar nebula. They may have functioned as tiny chondritic chemical reactors upon falling into the oceans.

Before reaching the Earth, organic molecules are exposed to ultraviolet radiation in interstellar space and in the Solar System. Experiments have now been carried out on amino acids exposed in Earth orbit in order to study their survival in space. The UV flux in the wavelength range less than 206 nm in the diffuse interstellar medium is about 100 million times weaker than in Earth orbit. This means that one week of irradiation in low Earth orbit corresponds to 275 000 years in the interstellar medium. Compared to ground experiments, space allows the exposure of samples to all space parameters simultaneously and the irradiation of a great number of samples under strictly identical conditions. Amino acids like those detected in the Murchison meteorite have been exposed to space conditions in Earth orbit onboard the unmanned Russian satellites, FOTON 8 and 11. Free exposed aspartic acid and glutamic acid were partially destroyed during exposure to solar UV. However, decomposition was prevented when the amino acids were embedded in clays, as has been reported by B. Barbier and others in 2002.

Amino acids and mini-proteins have also been subjected to solar radiation outside the MIR station for 97 days, both as solid films and embedded in a mineral material (montmorillonite clay, basalt powder and Allende meteorite powder). Different thicknesses of meteorite powder films were used to estimate the shielding threshold. After three months' exposure, about 50 % of the amino acids were destroyed in the absence of mineral shielding. The mini-proteins exhibited a noticeable sensitivity to space vacuum. Decarboxylation – the decomposition of the carboxylic acid groups -COOH with production of CO_2 – was found to be the main effect of decomposition by the solar UV. No polymerisation occurred and no conversion of L-amino acids into the D-amino acids was observed. Among the different minerals used as 5 μm films, meteoritic powder offered the best protection, whereas montmorillonite was less efficient. Significant protection from solar radiation was observed when the thickness of the meteorite mineral was 5 μm or greater. So we conclude that the harsh conditions of outer space, most notably ultraviolet radiation and vacuum, can be quite hostile and deadly to the building blocks of life, such as amino acids and proteins, when they are exposed without protection. However, it seems that considerable protection can be provided by quite moderate amounts of shielding materials such as clays and meteoritic powders. This is significant for the Litho-Panspermia hypothesis dealt with in Chapter 4.

We have seen that it is not very difficult to find sources for the basic building blocks of life, either on the primitive Earth or more likely having an extraterrestrial origin. We now ask: given these building blocks, can we reconstruct a primitive cellular life in the laboratory and, if so, what could that tell us about the likely mechanisms for the emergence of life on Earth?

If we consider living systems as they exist today, it seems reasonable to suppose that primitive life emerged as a cellular object, requiring boundary molecules able to isolate and protect the system from the aqueous environment (membrane), catalytic molecules to provide the basic chemical work of the cell (enzymes) and information retaining molecules that allow the storage and the transfer of the information needed for replication, such as RNA and DNA.

Primitive membranes may well have been formed by fatty acids composed of a hydrocarbon tail terminated by a carboxylic acid group -COOH. Fatty acids are known to form vesicles when the hydrocarbon chains contain more than ten carbon atoms. Such vesicle-forming fatty acids have been identified in the Murchison meteorite, as described by D. W. Deamer in 1998. However, the membranes obtained with these simple amphiphiles are not stable over a broad range of conditions. Stable neutral lipids made of two hydrocarbon chains attached to a glycerol molecule, CH_2OH-$CHOH$-CH_2OH, can be obtained by condensing fatty acids with glycerol or with glycerol phosphate, thus mimicking the stable contemporary phospholipids used in living systems, as we have already seen (Figure 2.7). G. Ourisson and Y. Yakatani have proposed that primitive membranes could initially also have been formed by simple isoprene derivatives.

Most of the chemical reactions in a living cell are carried out by enzymes or biological catalysts that are themselves proteins. Each type of protein is made from a particular combination of the 20 different homochiral L-amino acids used in life. As we have seen, the linkages between the amino acids are between the nitrogen atom in the NH_2 and the carbon atom in the COOH group and this link is called a peptide bond. Chains of amino acids linked together like this are therefore called peptides or, if there are many of them as in most proteins, polypeptides. So structurally we have:

AMINO ACIDS → PEPTIDES → PROTEINS

As we have said, the amino acids were probably available on the primitive Earth as complex mixtures. In order to construct the proteins from these amino acids some form of construction agent is needed. Now the number of condensing agents capable of assembling amino acids into peptides in water is limited, especially if we look for compounds that are likely to have been around in the prebiotic world. Some carbodiimides that have the general formula, R-N=C=N-R, can be used in water. The simplest carbodiimide, H-N=C=N-H, can be considered as a transposed form of cyanamide NH_2-CN, which is known to be present in the interstellar medium. In water, cyanamide forms a dimer, dicyandiamide, which is as active as carbodiimides in forming peptides. However, the reactions are very slow and do not produce anything larger than a 4-amino acid peptide.

Clays and salts can also be used to condense amino acids in water. In 1970, M. Paecht-Horowitz and others showed that amino acid adenylate anhydrides condense readily in the presence of montmorillonite, a clay. Mixtures of glycine, the most simple amino acid NH_2-CH_2-COOH, and kaolinite subjected to wet-dry and 25 °C – 94 °C temperature fluctuations, have been shown to produce so-called mini-proteins of up to five glycines. In 1996, J. P. Ferris and others reported the efficient mineral-catalysed (hydroxylapatite, illite clay) condensation of amino acids into long peptides. According to Nobel laureate C. de Duve, the first peptides appeared via thioesters. Thioesters are sulphur bearing organic compounds and presumably would have been present in the sulphur rich, volcanic environment of the early Earth. M. Bertrand, in 2001, described how they lead to short peptides in the presence of mineral surfaces.

S. W. Fox has shown that dry mixtures of amino acids polymerise when heated at 130 °C to give 'proteinoids'. The same can be achieved at 60 °C when polyphosphates are present. High molecular

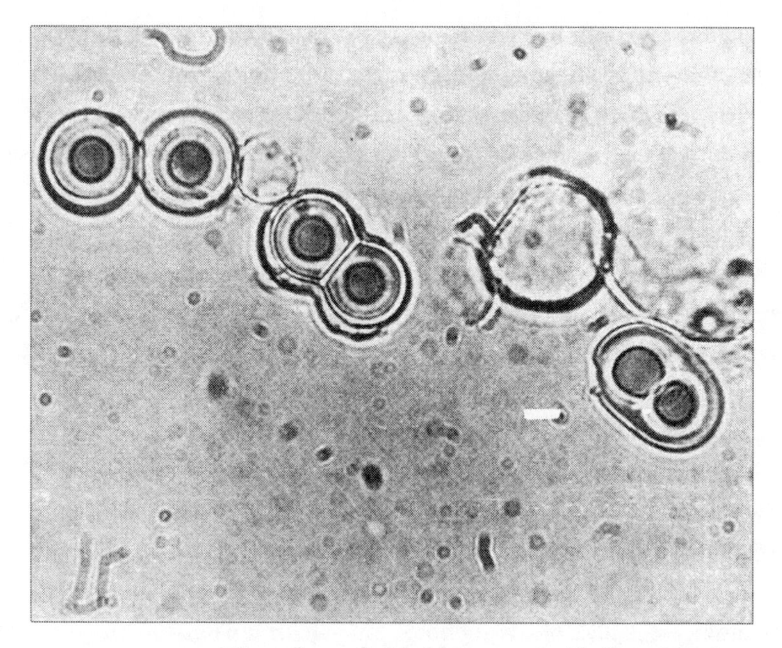

FIGURE 2.12 Microspheres obtained from proteinoids (from S. Fox).

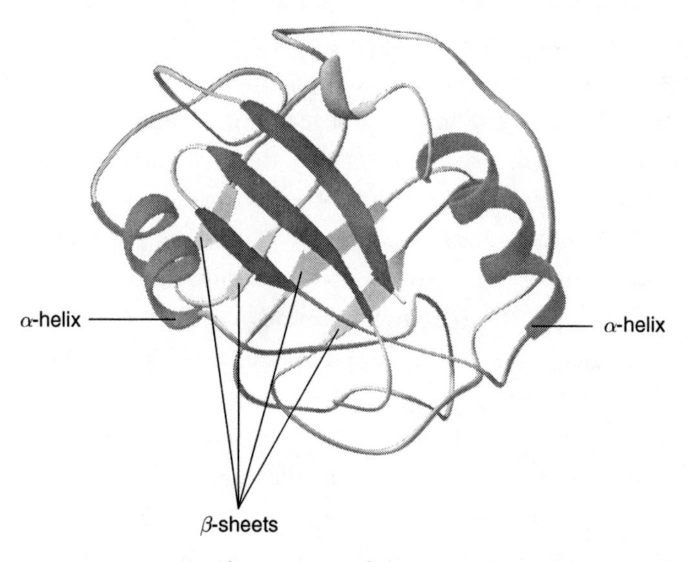

FIGURE 2.13 The structure of the protein cyclophilin, which contains both α-helices and β-sheets.

weights were obtained when an excess of acidic or basic amino acids were present. In aqueous solutions, the proteinoids aggregate spontaneously in microspheres of 1–2 μm, presenting a barrier resembling the biological membranes of living cells. The microspheres increase slowly in size from dissolved proteinoids and are sometimes able to bud and to divide. These microspheres are described as catalysing the decomposition of glucose. The main advantage of proteinoids is their organisation into particles, and they represent a dramatic increase in complexity (see Figure 2.12).

Chemical reactions capable of selectively condensing the protein amino acids at the expense of the non-protein ones have been found. Helical and sheet structures can be created with the aid of only two different amino acids, one hydrophobic or water repelling, the second hydrophilic or water attracting. Polypeptides with alternating hydrophobic and hydrophilic residues adopt a water-soluble β-sheet geometry because of hydrophobic side-chain clustering. Due to the formation of β-sheets, alternating sequences have a good resistance to chemical degradation. Interestingly, as A. Brack has shown in 1993, the aggregation of alternating sequences into β-sheets is possible only with homochiral (all-L or all-D) polypeptides. Short peptides have also been shown by B. Barbier and A. Brack to exhibit catalytic properties.

Turning now to the information carrying elements, we have seen how, in contemporary living systems, the hereditary memory is stored in nucleic acids built up with bases (purine and pyrimidine), sugars (ribose for RNA, deoxyribose for DNA) and phosphate groups. A. W. Schwartz, in 1998, has argued that the accumulation of significant quantities of natural RNA nucleotides does not appear as a plausible chemical event on the primitive Earth. Although purines are easily obtained from hydrogen cyanide or by submitting reduced gas mixtures to electric discharges, no pyrimidine synthesis from electric discharges has been published so far, and hydrogen cyanide yields only very small amounts of the base. Concerning the sugar, condensation of formaldehyde does lead to ribose, albeit among a large number of other sugars. The synthesis of purine nucleosides (the combination of purine and ribose) and of purine nucleotides have been achieved by heating the components in the solid state. The yields are very low and the reactions do not link the nucleotides exclusively through the natural 3′ and 5′ OH groups of the sugar, as in Figure 2.2. Interestingly, in 1996, J. P. Ferris reported the efficient clay-catalysed condensation of nucleotides into so-called mini-RNA (short chains of nucleotides) up to 55 monomers long.

The synthesis of mini-RNA is much more efficient in the presence of a preformed pyrimidine-rich RNA chain acting as a template. Non-enzymatic replication has been demonstrated by T. Inoue and L. E. Orgel. The preformed chains align the nucleotides by base-pairing to form helical structures which bring the reacting groups in close vicinity.

However, the prebiotic synthesis of the first short nucleotide chains remains an unsolved challenge.

Some years ago, it was found by A. J. Zaug and T. R. Cech that some RNAs, the ribozymes, have catalytic properties. In other words, they can also act as enzymes; for example, they increase the rate of hydrolysis of mini-RNA. They also act as polymerisation templates, since chains up to 30 monomers in length can be obtained starting from mini-RNA, 5 nucleotides long. Since this primary discovery, the range of catalytic abilities of these ribozymes has been considerably enlarged by directed test tube molecular evolution experiments, as reviewed by K. D. James and A. D. Ellington in 1995 and 1998. Since RNA can be at the same time the genetic material (the genotype) and the catalyst produced by the genotype (the phenotype), RNA has been proposed as the first living system on the primitive Earth (the so-called 'RNA world'). N. Ban and others in 2002 reported strong evidence for this proposal from the discovery that modern protein biosynthesis is catalysed by RNA. One should, however, remember, as stated above, that the synthesis of RNA itself under prebiotic conditions remains an unsolved challenge. It seems unlikely that life started with RNA molecules since these molecules are not simple enough. The RNA world appears to have been an episode in the evolution of life before the appearance of cellular microbes, rather than as the spontaneous birth of life.

(See Appendix 4 for more on RNA analogues and surrogates as clues to prebiotic chemistry.)

CONCLUSIONS

The year 2003 marked the 50th anniversary of the pioneering experiment of Stanley Miller. By demonstrating, in 1953, that it was possible to form amino acids – the building blocks of proteins – from methane, a simple organic molecule containing only one carbon atom, Miller generated a tremendous hope: would chemists be able to create in a test tube an artificial simplified life? What is the situation 50 years later? Although many great chemists have tackled the problem, it must be acknowledged that the dream has not yet been accomplished. By analogy with contemporary living cells, chemists tried to create a primitive cell, i.e boundary molecules (membranes), catalytic molecules (protein enzymes) and information molecules (nucleic acid, like RNA). Simple membranes and catalytic mini-proteins have been reconstructed in the laboratory but chemists failed to reproduce the formation of substantial amounts of RNA under prebiotic conditions.

The discovery of ribozymes that are able to carry the genetic information and to fulfil the catalytic activity allowed chemists to consider RNA as the start of life (the RNA world). However, this did not really help since the prebiotic synthesis of the primordial RNAs remains so far unsolved. The sugar ribose that is the main building block of the RNA backbone is unstable under the environmental conditions of the primitive Earth. To be available over a sufficiently long time, ribose must have been produced in large quantities via a chemical pathway for which we have no good model. The search for RNA surrogates is still in progress, but surrogates add additional steps in the overall scenario.

So far, chemists have followed the classical strategy used in organic chemistry, i.e. design → synthesis → test, a strategy promoted by Marcellin Berthelot and encapsulated in the phrase 'chemistry creates its own object'. So, over 50 years, this strategy has not been successful in this field. Perhaps this strategy was not best adapted to the problem, considering the complexity of the task at hand.

Another possible strategy would be to reconstitute the primeval soup with all sorts of minerals and organics, and to cook the soup for months. The modern methods of analysis, such as capillary electrophoresis and gas chromatography coupled with mass spectrometry, have gained such a degree of refinement they they should be able to detect any prebiotic-like organisation in the milieu. By doing so, chemists would have to let random chemistry replace their creativity. Some chemists are reluctant to let such a 'chance chemistry' replace their intelligence but this may be the price to be paid in order to achieve any further breakthroughs in this field.

We have seen, therefore, how Kauffmann's caveat represents a significant impediment to our understanding of the chemistry of how life arose on Earth. Nevertheless, significant progress has been made in identifying his 'bits and pieces of molecular machinery, patches of metabolic circuitry, genetic network circuitry and means of membrane biosynthesis'. Certainly enough to know how these elements or their remnants can be considered as chemical or biochemical markers of life present or past, and, in the light of that, what we should look for on other celestial bodies in the search for life.

3 The limits to life (water and extreme conditions)

Life on Earth is based on the chemistry of carbon in water. The temperature limits for the existence of life are given by the intrinsic properties of the chemical bonds involved in this type of chemistry at different temperatures. Two things are needed here. Firstly, the covalent bonds between carbon and the other atoms involved in the structure of biological molecules should be sufficiently stable to permit the assembly of large macromolecules that can perform catalytic and information-carrying functions or both. Secondly, non-covalent links (hydrogen and ionic bonds, Van der Waals interactions) should be relatively unstable. This is a very important point since only weak bonds can allow fast, specific and reversible interactions of biological molecules and macromolecules. These chemical constraints define the upper and lower temperatures for life, respectively. It is now known that terrestrial organisms can live in the temperature range from −12 °C to 113 °C (Figure 3.1).

HIGH TEMPERATURES

Presently, the maximum temperature limit known for terrestrial organisms is around 113 °C. For a long time the record was 110 °C, following the discovery of the microbe *Pyrodictium occultum* in shallow water near a beach of Vulcano Island in Italy, in 1982, by the group led by German biologist Karl Stetter. This record held for 15 years and was only recently surpassed, by the new record holder, *Pyrolobus fumarii*, a deep-sea microbe, again described by Stetter's group. The actual upper limit is, of course, unknown, but the barrier at 110–113 °C has so far survived the intensive search in deep-sea vents for organisms growing at higher temperatures. Many microbes living at temperatures close to or above the boiling point of water, the

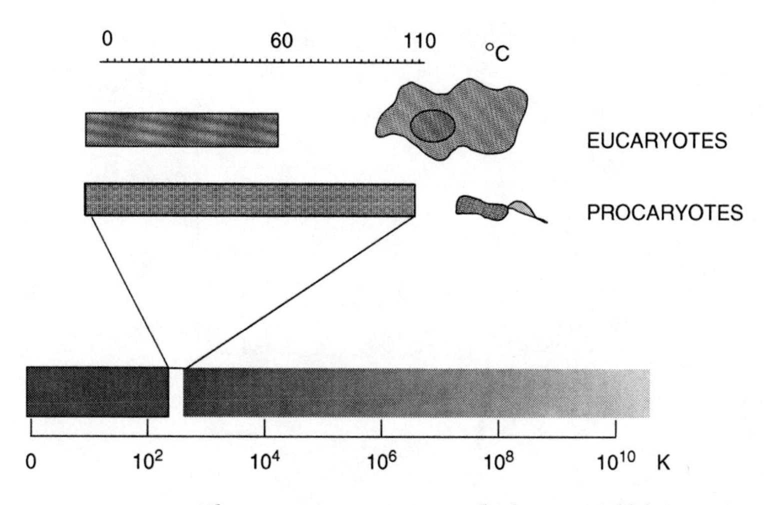

FIGURE 3.1 The temperature range over which terrestrial life can exist.

so-called hyperthermophiles, have been found in all places with volcanic activity, both on land and in the ocean. In particular, there are also large microbial populations at depth within hydrothermal vents, at ocean floor spreading centres, where they exploit the reduced chemicals in the hot vent fluid for energy and growth. They are all procaryotes, i.e. small unicellular micro-organisms without a nuclear membrane.

Procaryotes are divided into two phylogenetically distinct domains: Bacteria and Archaea (Figure 3.2). The most hyperthermophilic organisms at present all belong to the Archaea, the upper temperature limit for Bacteria being only 95 °C. Archaea living at 110–113 °C actually have the strongest growth at temperatures of 95–106 °C.

This all suggests that one or several critical factors prevent terrestrial life from proliferating efficiently even at 110 °C. This limiting factor cannot be the need for liquid water, because there are pressurised environments with liquid water at higher temperatures (the chimneys of hydrothermal vents), but they have been shown to be sterile by J. D. Trent and others. An important factor preventing life at temperatures well above 110 °C is the thermal instability of some

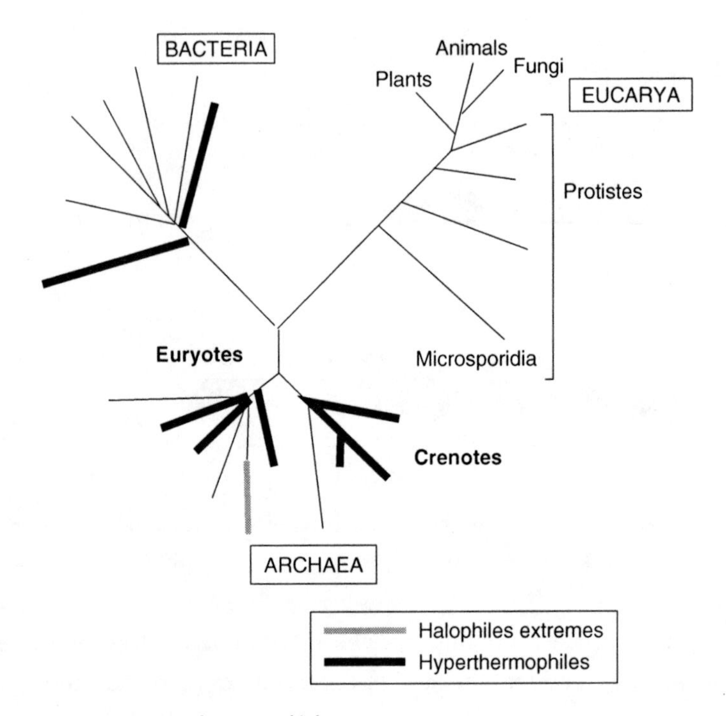

FIGURE 3.2 The 'tree of life'.

covalent bonds involved in biological molecules. Proteins can be very stable at high temperatures – some proteins from hyperthermophiles are still active in vitro after incubation at 140 °C. To achieve such stability, evolution has increased the number of non-covalent interactions between amino acids that maintain the folded structure of the protein.

The DNA double-helix is also very stable (at least up to 107 °C) as long as the two strands cannot freely rotate around each other, which corresponds to the intracellular situation as discussed by E. Marguet and P. Forterre. However, DNA is chemically degraded at high temperature by different mechanisms, the most important being the removal of purine bases and the subsequent cleavage of the strands at non-purine sites, as described by T. Lindahl. Accordingly, powerful repair mechanisms should exist in hyperthermophiles. In contrast to proteins and DNA, RNA is highly unstable at high temperatures because of the presence of a reactive -OH (hydroxyl) group in the 2′ position of the ribose part (see Figure 2.1) that can induce the cleavage of the strands. It has been estimated by French specialist Patrick Forterre, for example, that an RNA molecule of 2000 nucleotides (a typical length for a messenger RNA) should be cleaved in two pieces in about 2 seconds at 110 °C. Regions sensitive to heat-destruction appear to be protected in the transfer and ribosomal RNA of hyperthermophiles by methylation of the reactive hydroxyl group (-OH replaced by -OCH_3). The actual stability of RNA in vivo is, however, still unclear. This point is under study and is of utmost importance in the problem of life at high temperatures.

Many of the substances needed for metabolism and energy (metabolites) are also highly unstable at temperatures near the boiling point of water, in particular the ribonucleotides which contain energy-rich bonds, such as ATP (the energy supplier of the cell) or the substrates for DNA and RNA replication. Some energy-rich covalent bonds are already unstable in the temperature range typical of hyperthermophilic life (80–110 °C), but hyperthermophiles have successfully developed original strategies to bypass these limitations.

An important limiting factor preventing life above 110 °C could be related to membrane permeability. A. J. M. Driessen and others have shown that biological membranes become leaky at high temperature, allowing the free passage of ions which can lead to the breakdown of energy transfer in living organisms.

In conclusion, unless future work on subterranean organisms shows otherwise, either terrestrial life has been unable in the course of its history to design strategies for living at temperatures above 110 °C, or such a design is impossible because of the intrinsic limitations of carbon chemistry based life on Earth.

LOW TEMPERATURES

Life is extremely diverse in the ocean at temperatures of 2 °C. N. J. Russel has described how living organisms, especially micro-organisms, are also present in the frozen soils of arctic and alpine environments.

However, their best growth temperatures are usually well above the temperature of the site of isolation. Those organisms with optimal growth temperatures below 15 °C and minimal growth temperatures below 0 °C are called 'psychrophiles', while those capable of growth at 0 °C but with optimal growth temperatures above 15 °C are called 'psychrotrophs'. Psychrotrophs usually outnumber psychrophiles in a given biological environment, since they can benefit more efficiently from transient 'warm' conditions. For example, J. A. Nienow and E. I. Friedman have shown that bacteria living inside the rocks of dry valleys in Antarctica grow best at temperatures of 10–20 °C on rock faces exposed to the Sun in summer.

Cold-loving organisms are mainly bacteria (of all types) or eucaryotic micro-organisms. However, cold-loving archaea have been detected recently by E. F. DeLong and others in glacial water from Antarctica by direct amplification of their genetic material. Some of the most psychrophilic organisms are algae living in snow-covered areas, where they inhabit the upper 1 cm layer of snow. They often have optimal growth temperatures below 10 °C.

The lower temperature limit for life on Earth is not as clear as the upper limit, because it is very difficult to monitor growth and/or metabolic activity at sub-zero temperatures. As a consequence, very few fundamental studies have been carried out on this. N. J. Russel has also explained how the presumed lower limit for bacterial growth of −12 °C corresponds to the temperature at which ice is formed within cells. At lower temperatures, non-covalent bonds may become too strong for the kind of reversible reactions required for life to exist.

A major problem faced by cold-loving organisms at very low temperatures is the solidification of their membranes, risking the loss of the fluidity required for proper functioning. To overcome this, they usually change their lipid composition in order to increase membrane fluidity at low temperatures. Similarly, the structure of proteins from psychrophilic organisms is modified in order to increase their flexibility such that their activity is optimal at low temperatures compared to mid-temperature proteins. This type of modification is somehow a mirror of those occurring in hyperthermophilic proteins, i.e. while proteins from hyperthermophiles are usually more compact than their mid-temperature counterparts, homologous proteins from psychrophiles are often less compact, with more side-chains loosely connected to the protein core.

Terrestrial life finds it much easier to adapt to low temperatures than to very high temperatures. Whereas high temperature survival is restricted to certain groups of procaryotes, low temperature survival is widespread in procaryotes and eucaryotes. The reason for this difference is unknown, but it may be due to the breakage of covalent bonds, which is restricted to high temperature conditions.

HIGH-SALT ENVIRONMENTS

E. A. Galinski and B. J. Tyndall have described a well-studied group of extremophiles: the salt-loving organisms known as extreme halophiles. We know that salts are essential for terrestrial life (K^+, Na^+, Mg^{++}, Zn^{++}, Mn^{++}, Fe^{++}, Cl^-, etc.) because they are required as co-catalysts in many biochemical enzymatic activities. In that sense, all organisms are salt dependent. However, the salt concentrations which can be tolerated are usually quite low ($<0.5\%$) because high salt concentrations disturb the networks of ionic interactions that shape and hold together macromolecules. In the case of halophilic microorganisms (both eucaryotes and procaryotes), a wide range of salt concentrations (1–20 % NaCl) can be tolerated, and some procaryotes, the extreme halophiles, have managed to thrive in very salty environments such as salted lakes at up to 250–300 g per litre of NaCl. They are, in fact, so dependent on such high salt concentrations that they cannot grow (and may even die) at concentrations below 10 % NaCl.

Two strategies have been used by halophiles to cope with salty environments, where the main problem is to prevent the escape of water from the cell. The first strategy, used by eucaryotic algae and most bacteria, is to accumulate inside the cell (in their cytoplasm) small organic compounds, known as compatible solutes. In that case, the internal machinery of the cell is, in fact, protected from the extreme environment. The second strategy (mainly found in halophilic archaea) is to accumulate high concentrations of KCl and $MgCl_2$ in their cytoplasm (near saturation). In that case, the intracellular machinery is also in direct contact with the high salt content.

In halophilic archaea, the intracellular machinery is adapted to the high salt concentration of the cytoplasm, corresponding to very low water activity. Proteins from extreme halophiles are not only active at very high salt concentrations, but they are destroyed by the removal of salt. This can be dramatically demonstrated by the addition of water to a suspension of halophilic archaea observed under the microscope. The cells vanish as soon as the saltiness reaches a critical low point because membrane proteins dissociate when the salt concentration becomes too low. The mechanism of stabilisation of individual proteins at high salt levels is now beginning to emerge from structural studies. Halophilic proteins exhibit typical networks of negatively-charged amino acids at their surfaces, allowing them to retain a stabilising network of salt and water molecules.

ACIDIC AND ALKALINE ENVIRONMENTS

The chemistry of life on Earth is optimised for neutral pH but some micro-organisms have been able to adapt to extreme pH conditions, from pH 0 (extremely acidic) to pH 12.5 (extremely alkaline), at the same time maintaining the pH within the cell between pH 4 and 9. Many bacteria and archaea are acidophiles, living at pH values below 4, as reviewed by P. R. Norris and W. J. Ingledew. The record is held by the archaeon *Picrophilus oshimae,* which was found by C. Schleper and others and which grows optimally at pH 0.7 and still grows at pH 0. Like many other acidophiles, this organism is also thermophilic. Among acidophiles, the most thermophilic ones are Sulfolobales (archaea), which can grow at pH 2 and up to 90 °C. In many cases, these organisms are actually responsible for the acidity of the environment in which they live. For example, sulfolobales are sulphur-respiring organisms and produce sulphuric acid as a by-product of their metabolism.

Many bacteria and a few archaea, the alkaliphiles, live at the other extreme of the pH range, from pH 9 up to pH 12, as reviewed by W. D. Grant and K. Horikoshi. They are present everywhere on Earth. Some of them, which have been discovered in soda lakes rich in carbonates, are also halophiles (haloalkaliphiles). Most alkaliphiles are moderately thermophilic, but Karl Stetter and co-workers have recently described the first hyperthermophilic alkaliphile, *Thermococcus alkaliphilus*, as reported by M. Keller and others.

Both acidophiles and alkaliphiles rely on sophisticated transport mechanisms to maintain the pH within the cell near neutrality by absorbing or excreting protons. They have to be able to maintain adequate gradients of protons or sodium Na^+ to sustain their energy-producing machinery. Among them, the thermophiles also have to deal with the specific problem of maintaining the correct membrane permeability in an unbalanced ionic environment. This may explain why the upper temperature limits for acidophiles and alkaliphiles is presently 90 °C and not 110 °C.

HIGH-PRESSURE ENVIRONMENTS

As with temperature, the intracellular machinery cannot escape the influence of pressure. However, there are organisms in the deepest parts of the ocean at pressures of 1100 bar (atmospheres). The extreme pressure limit for life on Earth is unknown – environments of above 1100 atmospheres have not been explored. However, it might be quite high, because macromolecules and cellular constituents apparently only begin to break up at pressures of 4000 to 5000 atmospheres.

Some micro-organisms living in the deep ocean can be barophiles by choice, but most of them are only tolerant to high pressures. In fact, it is not yet clear if barophiles have had to invent specific devices to adapt to high pressure, as discussed by marine biologist Daniel Prieur. High pressure seems to increase the growth rate of some hyperthermophiles, but this phenomenon is not general and not very spectacular. G. Bernhardt and others have pointed out that although some proteins from hyperthermophiles are more active at high pressure, high pressure does not increase the thermal stability of macromolecules.

SUBTERRANEAN LIFE

For a long time, it was believed that there was no life in deep subterranean environments. It has now been recognised that bacteria (and probably archaea, too) actually thrive in the Earth's crust. Subterranean micro-organisms have been found by John Parkes and J. R. Maxwell and by T. O. Stevens and J. P. McKinley in subterranean oilfields or in the course of drilling experiments. For example, recent research conducted within the international Ocean Drilling Programme (ODP) has demonstrated that procaryotes are present much deeper in marine sediments than was previously thought possible, extending to at least 750 m below the sea floor, and probably much deeper. These microbial populations are substantial (e.g. 10 million cells in each cubic centimetre at 500 m below the sea floor) and likely to be widespread. H. Furnes and others have shown that microbes can alter volcanic glass to depths of at least 432 m. This glass comprises a substantial volume of the volcanic component of the ocean crust and may have significance for chemical exchange between the oceanic crust and ocean water. These data provide a preliminary, and probably conservative, estimate of the biomass in this important new ecosystem: about 10 % of the surface biosphere. These discoveries have radically changed our perception of marine sediments and indicate the presence of a largely unexplored deep bacterial biosphere that may even rival the Earth's surface biosphere in size and diversity!

High numbers of procaryotes have also been found in association with subterranean gas sources. This, together with the presence of large microbial populations in oil reservoir fluids, as described by Karl Stetter and others, such as H. Haridon, F. P. Bernard and P. Rueter, suggests that deep bacterial processes may even be involved in oil and gas formation. Procaryotes are also likely to drive deep geochemical reactions such as mineral formation and dissolution, be responsible for the magnetic record via the production of magnetic minerals and produce deep methane gas essential for gas source formation.

FIGURE 3.3 A deep-sea oil rig. Microbes exist at up to 1km deep in these wells. (Courtesy Shell.)

Procaryotes in the deep biosphere must be uniquely adapted to live in this extreme environment, and populations have even been shown to increase at depths greater than 200 metres as they exploit deep geochemical fluxes, such as geothermal methane and salt water incursions.

T. O. Stevens and J. P. McKinley also obtained evidence for the existence of anaerobic subsurface microbial systems in basalt aquifers. These communities apparently get their energy from geochemically produced hydrogen. Because the energy sources and inorganic nutrients are both supplied by geochemical means, such a subsurface microbial ecosystem could a priori persist indefinitely, even if the conditions on the Earth's surface became alien to life.

Since the temperature increases with depth, it has been suggested by T. Gold that hyperthermophiles are abundant in subterranean environments, forming a so-called 'deep hot biosphere'. Indeed, procaryotic activity has been reported at up to 120 °C and possibly even higher, by B. A. Cragg and R. J. Parkes, i.e. above the upper temperature limit found in the laboratory.

The existence of these deep communities clearly suggests that life may also exist deep below the surfaces of other planets. So the conditions for life may actually improve below the surface, for example, on Mars. The absence of surface life, therefore, may not necessarily indicate the absence of all life and so the search for life should not be limited to a planet's surface.

The revival of micro-organisms taken from ancient rocks, salt and coal has been reported many times. There are several claims of micro-organisms being revived from rocks more than 100 million years old, some even from Precambrian rocks 650 million years old. This has all been reviewed by M. J. Kennedy. These claims have usually been disputed on various grounds, such as contamination or theoretical impossibility, but Kennedy concluded that the observations are too many to be dismissed en bloc. In particular, the US Department of Energy has established a collection of 5000 revived micro-organisms, such as bacteria and fungi from various subsurface sites about 200 million years old. For Kennedy the best alternative to revival is *in-situ* reproduction, which is as interesting as real revival, particularly from the perspective of recovering ancient living forms on Mars or other planets.

Apart from the availability of resources, the balance between latent life and *in-situ* reproduction in subsurface conditions should be dependent on temperature effects: *in-situ* reproduction being favoured at high temperatures and latency at low temperatures. Micro-organisms in general, and even some macro-organisms, are extremely resistant to freezing temperatures. It is well known that micro-organisms are routinely kept alive for years in liquid nitrogen. Indeed, at extremely low temperatures, most deleterious processes that can impair survival are slowed down.

Many of the micro-organisms revived from ancient material are spore-formers. Bacteria of the Gram-positive kingdom (one out of the roughly 12 bacterial kingdoms), have indeed developed the ability to produce extraordinarily resistant spores that can survive extremely harsh conditions (high temperature, absence of nutrients, high doses of radiation). Spores are also produced by fungi and minute plant seeds, and protozoan cysts can be considered as types of spores.

The question of how long spores can survive depends critically on the stability of DNA molecules over very long periods, and the possibility for the organism to repair any lesions suffered during the dormant stage, once the conditions again become favourable. T. Lindahl has discussed how, even in the absence of UV or ionising radiation, DNA is subjected to spontaneous chemical modifications to key parts of its structure. These reactions are very slow at low temperatures but can still produce significant damage in the very long term. This is important for the possibility of revival of organisms that are several million years old.

DNA in spores is protected by specific DNA binding proteins that prevent depurination and probably other damage, but it is not clear how effective this protection can be over very long periods. As has been stated above and discussed by E. Marguet and P. Forterre, high salt concentrations protect DNA against chemical modifications. Thus, halophilic organisms might have more chance to survive, and/or for their DNA to remain intact, than non-halophilic ones. Therefore, frozen extreme halophiles may have the best chances for long-term survival.

Natural radioactivity in rock might be also a problem for long-term DNA stability. However, some micro-organisms are extremely radiation-resistant. This can be a secondary adaptation to desiccation that produces DNA lesions, leading to double-stranded breaks when the cells are again in contact with water. M. D. Smith has shown that the radiation-resistant bacterium *Deinococcus radiodurans* exhibits a very efficient repair-recombination mechanism that allows the cell to reconstruct intact chromosomes from damaged ones.

IMPLICATIONS FOR EXOBIOLOGY IN FUTURE SEARCHES

What lessons do extremophiles and survival in the extreme teach us about possible extraterrestrial environments and our chances of identifying relic or active extraterrestrial life in the Solar System? We should first make the distinction between the conditions required for the emergence of life and those for its evolution and maintenance.

In some theoretical scenarios, life appeared at very high temperatures. Karl Stetter proposed that today's hyperthermophiles can be viewed as relics of the last common universal ancestor of all living beings. In that case, the origin of life would have required the presence of stable warm biological environments at an early stage of planetary evolution. However, this 'hot origin of life' hypothesis has been challenged by Patrick Forterre, based on the idea that early life on Earth was probably dominated by RNA instead of DNA, and RNA is very unstable at high temperatures.

The 'hot origin of life' scenario is mainly linked to the grouping of hyperthermophiles at the base of the universal 'tree of life', on each side of its root, as deduced from rRNA analysis by Stetter. However, the rooting of the universal tree has been disputed by Forterre, and recent data indicate that rRNA phylogenies can be very misleading. In particular, it is now clear that microsporidia (eucaryotes without mitochondria), which were supposed to be the most primitive eucaryotes according to rRNA phylogeny, are in fact fungi, as referenced by J. D. Palmer.

Even if life did not originate at very high temperatures, the production of efficient catalysts at freezing temperatures might have been an impossible task at an early stage of evolution. The most attractive hypothesis might be that life appeared in a moderately thermophilic environment: hot enough to boost catalytic reactions, but cold enough to avoid the problem of macromolecules breaking up due to high temperatures.

The study of micro-organisms living in subterranean environments can teach us much about the effect of pressure on living organisms. This is of great importance, as many possible biological environments in the Solar System experience either very low or very high pressures. The study of subterranean biological environments may also be relevant to the problem of the origin of life in general. It may have been that life originally developed on Earth below the surface and then grew upwards when conditions became less extreme in the early Archean period. If this is correct, life might have appeared on several other Solar System bodies in the appropriate temperature zone at the boundary between external cold and internal heat.

What can we learn from the biology of terrestrial extremophiles? An important observation is that most terrestrial extremophiles are procaryotes. This is especially striking in the case of thermophiles and hyperthermophiles, since all micro-organisms living at 60–110 °C are procaryotes. The reason for this discrimination is unknown. It could be related to the requirement for long-lived messenger RNA in eucaryotes, as described by Patrick Forterre, but this is still only a hypothesis. Why only archaea thrive above 95 °C is also still a mystery. More work is needed to understand these different frontiers. On the other hand, all terrestrial present-day extremophiles, although procaryotes, are complex micro-organisms, the products of long evolution, which exhibit elaborate mechanisms for coping with the extreme environments that they find themselves in. For example, terrestrial hyperthermophiles have developed strategies to protect their macromolecules, such as proteins, DNA and RNA, against the damaging effects of very high temperatures, a feature also proposed by Forterre. Similarly, the two strategies used by halophiles to cope with salty environments, i.e. the production of compatible solutes and the accumulation of high intracellular salt concentrations, require the presence of very effective transport systems across membranes and/or complex metabolic pathways to synthesise compatible solutes, two hallmarks of highly evolved organisms. This again suggests a priori that life, as we know it, probably evolved first in a mild environment before invading more extreme ones.

The terrestrial dichotomy between procaryotes and eucaryotes raises other interesting questions. For example, would all life forms on other planets automatically extend at least up to 110 °C or is this expansion dependent on the appearance of a procaryotic lifestyle? The answer to this question requires us to define a priori what the nature of a procaryote is. Traditionally, procaryotes are considered to be ancient and primitive forms of life on Earth. In such a scenario, the procaryotic type of cellular organisation is often viewed as an early stage of life's evolution, which should have been common to all life, consequently making it the most widespread life form in the Universe. However, some scientists, such as D. C. Reanney and P. Forterre, think that procaryotes are very evolved organisms that originated by reduction from more complex forms of life. In that case, the presence of life in many extremophilic environments could be dependent on the historical appearance of 'procaryotes'. It might be important to determine which is correct in order to define more precisely what we are looking for on other planets. In any case, the procaryote/eucaryote dichotomy may not be valid for all possible forms of life in the Universe. Simpler, different or more complex forms could exist.

It has been suggested that life might have appeared on planets only within a narrow spectrum of physical parametres and that, once life existed, these requirements could be relaxed (but not too much) for evolved life forms to proliferate. However, if these conditions changed drastically in some direction, such as freezing and desiccation, living organisms could survive, waiting for better times for very long periods, possibly on an astronomical scale. Question marks predominate in this area of research. Some of them have been reviewed by Kennedy and others. How long can extremophiles or resistant forms survive in hostile conditions? What is the migration rate of micro-organisms through a rock structure? For how long can nucleic acids survive long-term storage? How can new methods to assess the antiquity of 'revived' terrestrial micro-organisms be designed? How can bacterial reproduction in very old samples be studied?

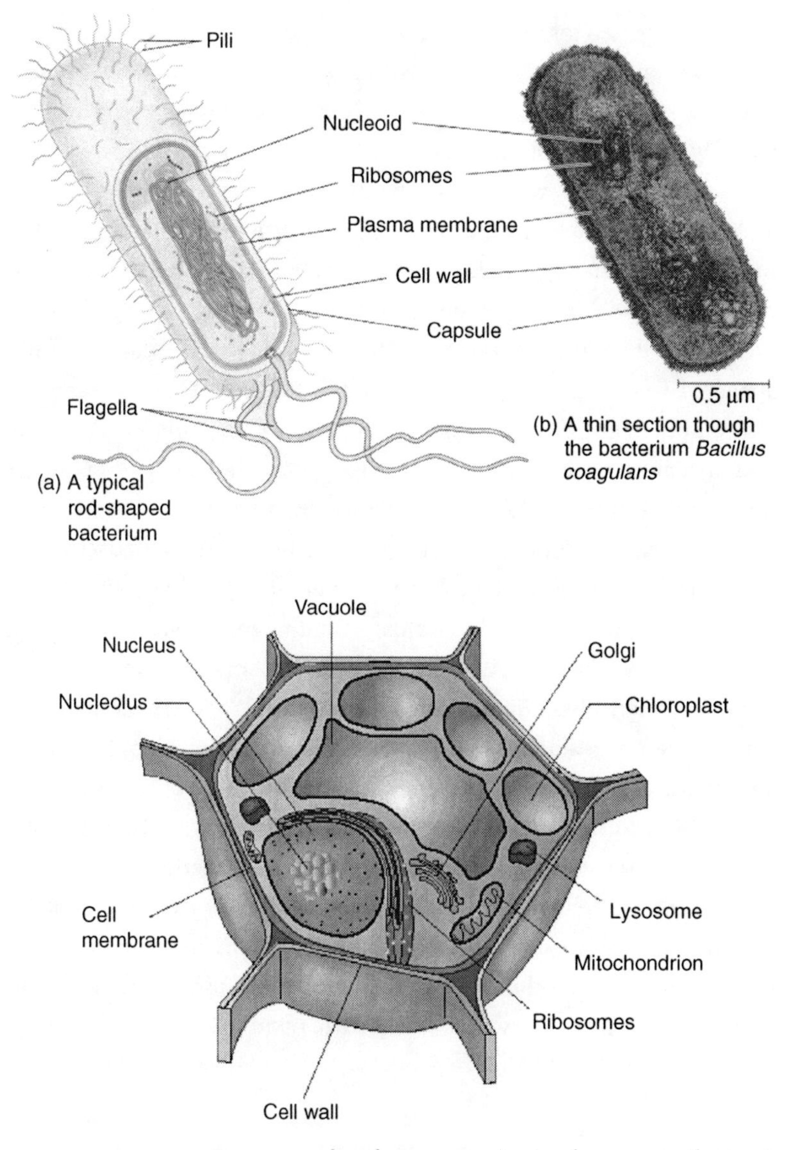

FIGURE 3.4 Some examples of procaryotes (top) and eucaryotes (bottom).

In general terms of survival, spores immediately come to mind, but sporulation is not a general phenomenon of the living, but an historical evolutionary invention that may or may not have occurred in the history of life on other planets. Furthermore, terrestrial micro-organisms can be kept frozen for years and, though they are not spores, they survive. As seen from the above, the discussion usually centres around procaryotes, but some eucaryotes (acaria) could exhibit surprising and at present scarcely explored resistant properties.

In conclusion, recent discoveries demonstrate that bacteria and archaea have conquered the whole planet, thus invading all possible niches where there is liquid water, even episodically. In fact, these micro-organisms dominate terrestrial life. The supremacy of these rather simple living systems is illustrated by their duration (they appeared first and will probably last the longest), their remarkable resistance to extreme conditions, their ubiquity (the only sterile places on Earth are probably incandescent lavae), their large taxonomy (each division of Bacteria and Archaea is greater than the totality of the three multicellular kingdoms: plants, fungi and animals) and their usefulness in generating part of the Earth's atmospheric oxygen and contributing via symbiosis to the other life forms, including humans.

Biologists have shown that micro-organisms can survive under extreme conditions. Life has continued to develop very well in water that is either very acidic, alkaline, or a strong brine solution. It has also survived and flourished in water at high pressure and at temperatures up to 113°C. A flourishing biosphere has been discovered a kilometre below the surface of the Earth. Procaryotes, i.e. bacteria- and archaea-type micro-organisms, constitute therefore the priority target in our search for extraterrestrial life. Habitable zones will be explored where liquid water is thought to be present under conditions that can be quite far away from Darwin's 'warm little pond'.

4 The transfer of life between planets

It's 27 December 1984. The dim light of the midnight Sun reveals a waste landscape, unchanged for tens of millions of years. The landscape is silent, deserted and bare, covered by perpetual snow and sitting on a 4-km thick layer of ice. The sound of an engine disturbs the eerie silence. A snowmobile approaches, stops, and the driver, a member of the US meteorite hunting expedition in the Allan Hills region of the Far Western Icefield of Antarctica, descends, and picks up a dark stone. It is about 15 centimetres in diameter, and it stands out against the monotonous blanket of snow and ice. It's a messenger from space. 'Yowza-Yowza' is written in the field notes about the find. There follows the dry technical description of the rock: 'Sample number 1539, highly shocked, greyish-green, achondrite, 90 % covered with a fusion crust.' Since it was the most unusual rock collected during that field season, it was the first specimen to be processed and analysed. Its description, ALH84001, reflects the sampling site (ALH for Allan Hills), the year of collection (1984) and the order of processing (001 for the first rock studied). Nobody in 1984 realised that this rock would much later cause a fierce discussion on the likelihood of life on Mars and the possibility of a viable transport mechanism for the exchange of life between Mars and the Earth.

It took nearly 10 years (until 1993) for careful analyses to convince people that ALH84001 was a Martian meteorite. R. N. Clayton and T. K. Mayeda have explained how the oxygen isotope composition shows that Martian meteorites are different from Earth rocks, rocks from the Moon or other meteorites that mainly originate from the asteroid belt. In addition, a considerable amount of data from petrological, geochemical and noble gas analyses can be used to determine the origins of a meteorite.

FIGURE 4.1 Meteorite ALH84001 as found in Antarctica. The cube is 1 centimetre. (Courtesy NASA.)

ALH84001 is the oldest among the 28 meteorites so far found on Earth that have clear signs of originating from Mars. They are called SNC meteorites, an acronym based on where the earliest such meteorites were found: the meteorite Shergotty was found near the town of Shergahti, India, in 1865; Nakhla was picked up near the village of El-Nakhla, Egypt, in 1911; and Chassigny near the village of Chassigny, France, in 1815.

The history of the meteorites is imprinted in their mineralogy and geochemistry. As we have seen, the stable oxygen composition testifies to Mars as the home planet of the SNC meteorites. ALH84001 was crystallised from a lava flow very early in Martian history, about 4.5 billion years ago. It is a sample of the early Martian crust. Hence, it experienced the whole history of the Martian climate, from the warm and humid environment believed to have prevailed during the first

FIGURE 4.2 Impact of a meteorite on a planet. (© Don Davis.)

several hundred million years of Martian history, to an era of gradual drying, cooling and thinning of the atmosphere.

Then, about 16 million years ago, a catastrophic event occurred on Mars: the impact of a huge meteorite left a crater over 30 km in diameter. A large amount of material was ejected into space, including the piece of rock we now know as ALH84001.

Once blasted into space by this huge impact, ALH84001 spent approximately 16 million years in space, as inferred from its cosmic exposure age, a lonely wanderer between the planets. It required several collisions with other ejecta or small bodies – some of these collisions may have even fragmented the initially expelled rock – until the meteorite was pushed from the Martian orbit into an Earth-crossing orbit. Finally, it was drawn to Earth by Earth's gravitational field and entered the atmosphere. Its outermost layers were heavily heated, but because the fall through the atmosphere lasted only a few tenths of a second, these layers served as a kind of heat shield, and the heat did not have time to penetrate into the interior. Only the surface melted to glass, as indicated by the fusion crust covering 90 % of the meteorite. From the density of nuclear tracks produced by cosmic rays it was concluded that about 5 cm of the outer crust were shed during the entry into the Earth's atmosphere.

ALH84001 fell onto the icy surface of Antarctica. This blue ice area of Antarctica does not particularly attract meteorites. Meteorites fall all over the globe. However, they are easier to spot against the light-coloured background of the ice rather than sitting on rock or lying embedded in soil. More than 10 000 meteorite specimens have been recovered in this way since a co-ordinated search was started in 1976. Over thousands of years, the katabatic winds that roar down the ice cap from the South Pole to the ocean have scoured away the ice, leaving behind the meteorites that have been sprinkled throughout the volume of the ice. It took about 13 000 years in the ice before ALH84001 was picked up and entered into the meteorite collection at the Johnson Space Center (JSC) of NASA in Houston.

ALH84001 became famous in 1996 when David McKay and his research group at JSC announced that this enigmatic rock appeared to disclose an amazing secret: evidence for primitive life on early Mars. However, the difficulties in reaching a clear-cut interpretation were reflected in their paper in the journal *Science*: while *'none of the observations is in itself conclusive for the existence of past life'* (on Mars), that when considered collectively, particularly in view of their spatial association, *'they provide evidence for primitive life on early Mars'*. In the years following this controversial report, ALH84001 became one of the best investigated meteorites; but the dispute still continues today as to whether or not ALH84001 is a messenger of past Martian life.

The discovery of Martian as well as lunar meteorites that have landed intact on the surface of the Earth throughout its history gives support to the idea that our planet Earth is not an isolated body within the Solar System. The exchange of material between the planets happened in the past, probably quite frequently. This has sparked renewed interest in the question of whether or not viable micro-organisms

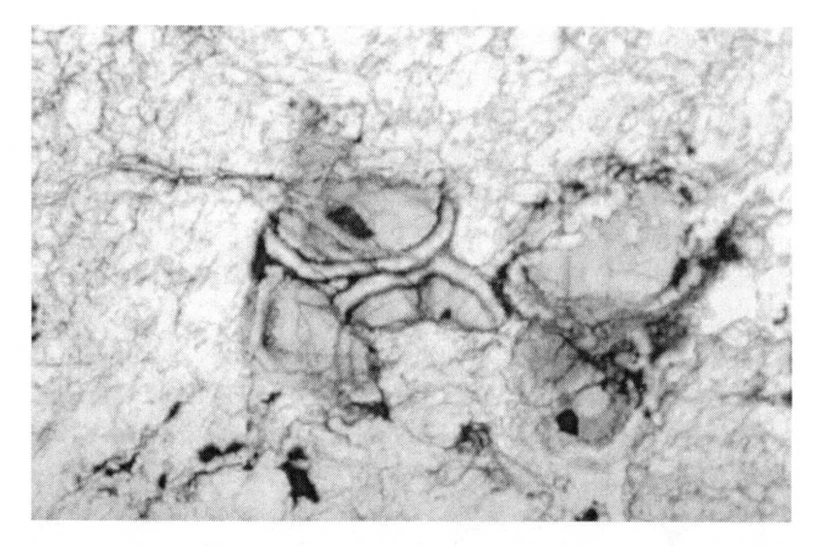

FIGURE 4.3 Carbonate globules in ALH84001. (Photo by Allan Treiman.)

embedded within impact ejecta could be transported between planets, as has been proposed by C. Mileikowsky and others. If so, this could have had a profound effect on evolution, and would make the discovery of life elsewhere in the Solar System that much more difficult to interpret.

The conception of life being distributed through space is not a new one. From the nineteenth century, and into the beginning of the last century, the theory of Panspermia has been discussed among scientists. The theory postulates that microscopic forms of life, for example spores, could be transported through space, driven by the radiation pressure from the Sun, and thereby seeding life from one planet to another, or even between planets of different solar systems. In 1903, the Nobel laureate for chemistry, Svante Arrhenius (1859–1927), laid the foundations of Panspermia in his *Die Umschau* article 'Die Verbreitung des Lebens im Weltenraum' (the distribution of life in space). Arrhenius thought that small particles like spores would be powerfully pushed through space by this radiation. Clearly visible celestial examples of the Sun's radiation pressure are the tails of active comets that stream away from the Sun.

However, in the vicinity of the Earth, the radiation pressure of the solar light is quite low (0.4×10^{-10} bar) compared to other forces in space, e.g. accelerations resulting from collisions with other bodies in space. Because its effectiveness decreases with increasing size of the particle, this mechanism holds for very tiny particles only, such as single bacterial spores (about one thousandth of a mm in diameter). Furthermore, the radiation pressure decreases with the square of the distance from the Sun.

Arrhenius estimated that single spores could be driven into interstellar space with enough speed to reach another star within a 'reasonable' time, which he considered to be about 3000 years. He was not aware that single micro-organisms are killed within seconds when exposed to the harmful solar UV radiation in space. Their high sensitivity to extraterrestrial UV radiation, even if in the more resistant dormant state of bacterial spores, was first demonstrated in laboratory experiments simulating space conditions, and later, in experiments performed directly in outer space, and reported by G. Horneck in 1993. Furthermore, he ignored the reverse effect of the radiation pressure of the target star, which would prevent small particles from entering the region around that star.

An alternative mechanism for the transfer of life between planets was suggested by the British physicist Lord Kelvin (1824–1904), a contemporary of Arrhenius. He favoured the rocky version of Panspermia, with fragments of extraterrestrial rocks carrying microbes as blind passengers within cracks and transporting life from one planet to the other. However, at that time, no mechanism was known for accelerating rocks to escape velocities in order to leave their planet of origin. So, the idea was discarded in favour of the classical Panspermia theory.

It is interesting to note that quite a large fraction of scientists at the turn of the nineteenth century, like Arrhenius, believed that life is eternal and that Panspermia might therefore be a plausible process for distributing life throughout the Universe. They came to this conclusion mainly on the basis of two observations made at that time.

Firstly, the failure of experiments attempting to reproduce the processes of the origin of life on Earth – which have still not succeeded to the present day – and secondly, Pasteur's convincing experiment disproving 'Urzeugung'– the spontaneous generation of life.

In 1871, however, Charles Darwin (1809–1882) had already proposed an alternative pathway for the appearance of life on Earth. Darwin proposed the *in-situ* production of complex organic compounds on Earth as prerequisites for the origin of life on Earth. 'If we could conceive in some warm little pond, with all sorts of ammonia and phosphoric salts, light, heat, electricity etc. present, that a protein compound was chemically formed . . . ' More than 50 years later, A. I. Oparin and J. B. S. Haldane independently developed a scenario for the origin of life from inanimate matter, but it took about 80 years before this idea of a 'warm little pond' was tested experimentally. Stanley Miller and Harold Urey exposed a mixture of methane, ammonia, hydrogen, and water to electrical discharges in a simulation of the supposed atmosphere and oceans of the early Earth. In 1953, Miller identified, among the compounds formed, amino acids, the building blocks of proteins. Since then, a variety of possible atmospheres ranging from highly reducing (e.g. hydrogen, ammonia, methane, water) to non-reducing compositions (e.g. carbon dioxide, nitrogen, water) have been exposed to different energy sources to determine their potential for abiotic organic synthesis. Non-reducing atmospheres result in a very poor production of amino acids and other nitrogen-containing compounds, if they are formed at all. However, just such a non-reducing atmosphere is now supposed to have prevailed on the early Earth. Therefore, *in-situ* abiotic synthesis of amino acids on the early Earth would appear to have been a rather rare and unlikely process. Today, the most favoured idea is that the import of organic compounds from extraterrestrial sources triggered the development of life on Earth, which then evolved in liquid water, as described by two of these authors in 1992 (G. Horneck and A. Brack) and which has been covered in more detail in Chapter 2.

Since then, several possible pathways for the abiotic synthesis of organic compounds have been identified as quite common processes, either on the primitive Earth (e.g. in some protected niches of reducing gas mixtures) or elsewhere in the Universe. At that point, Panspermia was not considered further as a likely or necessary mechanism for the appearance of life on Earth. The main arguments against it were that it cannot be experimentally tested and that even the very resistant bacterial spores would not survive long-time exposure to the hostile environment of space, especially to vacuum and radiation. Furthermore, it was argued that Panspermia does not help us in our efforts to understand the origin of life, because it shifts the problem elsewhere in the Universe without solving it.

To counteract some of these objections, it has been suggested that Panspermia can also be 'directed' by other intelligent beings from another planet – or another solar system – who purposely send protected packages of special microbes on interstellar missions. This 'directed Panspermia' was proposed by Nobel laureate Francis Crick in 1981. Crick had received the Nobel Prize for discovering the double-helix structure of DNA with James Watson and Maurice Wilkins 50 years ago. The theory is mainly based on the argument that the history of all life on Earth goes through a bottleneck called the last universal common ancestor, and that the genetic code is universal for all living beings of our biosphere. Although Crick agreed that the arguments he employed in favour of 'directed Panspermia' are somewhat sketchy, the possibility of life having reached the Earth via 'infection' should not be ruled out as one of many possibilities. Although this 'directed Panspermia' has not found many other supporters, the concepts of terraforming Mars, i.e. to intentionally modify Mars in such a way as to make it more habitable and eventually more Earth-like, as has been proposed by Chris McKay and others in 1992 and is described in Appendix 8, indicate the technical feasibility of 'directed Panspermia', at least within our own Solar System.

The concept of terraforming Mars foresees two steps. In the first, changes of the environmental parameters of the planet by planetary engineering are proposed, such as the distribution of volatiles and the modification of parameters like surface temperature and pressure, the atmospheric composition and opacity, planetary reflection or albedo, precipitation and humidity. This would warm the climate sufficiently to allow the presence of liquid water and increase the thickness of the atmosphere by the release of frozen gases. The second step would involve biological engineering, i.e. the production of genetically engineered micro-organisms and their implantation as pioneering microbial communities able to proliferate in the newly clement, though still anaerobic, Martian environment. For terraforming Mars, e.g. for the production of a breathable atmosphere, timescales of the order of 10 000 years are foreseen.

However, one has to note that the purposeful introduction of terrestrial life forms on Mars or any other planet or moon of our Solar System by means of orbiters, entry probes or landers, and their possible proliferation in the new environment, would entirely destroy the opportunity to examine the planets or moons in their pristine condition. In order to prevent the introduction of microbes from the Earth to another celestial body or vice versa, whether this occurs intentionally or unintentionally, the concept of planetary protection has been developed. The intention of this is twofold: firstly, to protect the planet being explored and to prevent jeopardising search-for-life studies, including those for precursors and remnants; and, secondly, to protect the Earth from the potential hazards posed by extraterrestrial matter carried on a spacecraft returning from another celestial body. Planetary protection issues are bound by an international treaty of the United Nations of 1966 and a follow-on agreement of 1979. Based on these treaties, a concept for contamination control has been elaborated by the Committee of Space Research (COSPAR) taking account of specific classes of mission and target combinations. This is recommended to be followed by all space-faring organisations.

The discovery of the 28 Martian meteorites has led to a revisitation of Panspermia. The impact scenario allows even boulder-sized rocks to reach escape velocities, propelled by the shock waves that are released when a km-sized meteorite hits a planet. Material located close to the surface and at the rim of the impact crater could even be

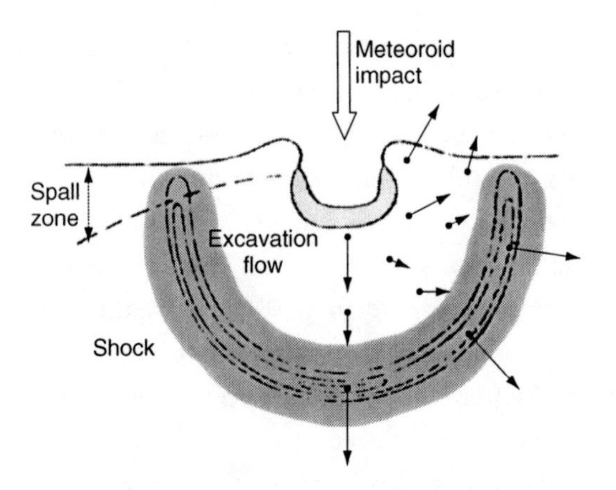

FIGURE 4.4 The spallation process whereby rocks escape without severe heating. (Courtesy H. J. Melosh.)

ejected at quite moderate temperatures, below 100 °C, thanks to the so-called spallation process as shown in Figure 4.4. To escape the gravity field of Mars, a body needs a velocity of 5 km/s. But this needs a high shock pressure of the impact event, so that rocks ejected with a velocity of 5 km/s or more should be shock molten or even vaporised. It is difficult to see how life forms could survive this. However, some of the Martian meteorites are only moderately shocked. In 1992, A. Bischoff and D. Stöffler estimated post-shock temperatures ranging between 100 °C and 600 °C. Indeed, in 2000, B. P. Weiss and others claimed that some data suggest even lower temperatures for ALH84001, probably not above 40 °C.

A way out of this dilemma can be envisaged based on a mechanism originally proposed by H. J. Melosh in 1985. A small amount of material may be thrown from the surface into space by stress-wave interference in the vicinity of a large impact, where the resulting shock is considerably reduced by a superimposition of the reflected shock wave on the direct one (Figure 4.4). Due to this spallation effect, moderately shocked fragments from the uppermost layer of the target can be accelerated to very high velocities, such as the escape velocity of Mars, or to even higher velocities. Hence, although impacts are very cataclysmic, a certain fraction of the rocks ejected are not heated above 100 °C.

These impact mechanisms would work not only on Mars, but also on the much larger Earth, whose escape velocity is 11.2 km/s, more than twice that of Mars. Thus the impact mechanism allows material to flow in both directions, from Mars to Earth, but also from Earth to Mars, although the latter may be less frequent.

It has been calculated that, within the last 4 billion years, more than 1 billion fragments of 2 m in diameter or larger were ejected from Mars from spall zones where the temperatures did not exceed 100 °C. When they leave their home planet, most fragments orbit the Sun, usually for timescales of a few hundred thousand or perhaps several million years, until they either impact another celestial body or are expelled from the Solar System. It has been calculated that for travel times not exceeding 8 million years – this number corresponds to the travel times observed for most of the Martian meteorites – about 5 % of the low temperature, only slightly shocked Martian meteorites have finally been attracted by the gravitational forces of the Earth. Because of the large number of ejecta, this number of Martian meteorites arriving on Earth is still high, of the order of fifty million. The 28 Martian meteorites so far found on Earth are clearly a tiny fraction of these.

Because the escape velocity of the Earth is more than twice that of Mars, a ten times smaller number of moderately shocked ejecta from the Earth is probable. It is calculated that about 0.1 % of them would arrive on Mars within 8 million years. This still gives a sizeable number of about a hundred thousand meteorites arriving on Mars from the Earth.

This revised Panspermia theory, using the impact scenario, should be able to answer whether or not this impact scenario would be suitable for transferring life forms from their home planet to a neighbouring planet. For this to happen, a number of barriers need to be overcome: firstly, the original impact site must happen on an inhabited planet at a location where living organisms, preferably micro-organisms, are present. Secondly, the material ejected from the spall zone of the impact crater with velocities sufficiently high to allow escape must contain living (micro)organisms. Thirdly, the (micro)organisms enclosed in the ejecta must survive the ejection process, as well as the extended periods of travelling in interplanetary space. Fourthly, when the meteorite arrives at the other planet, the (micro)organisms must cope with the landing process, which might include entry through an atmosphere, as well as thermal and pressure shocks during impact on the surface. Lastly, the new planet must provide a habitable environment accessible to the invading (micro)organisms.

In 2001, Ben Clark of Lockheed Martin Aeronautics developed a formula to include the uncertainties connected with a viable transfer of life within our Solar System, as follows:

$$P_{AB} = P_{biz}.P_{ee}.P_{sl}.P_{ss}.P_{se}.P_{si}P_{rel}.P_{st}.P_{sp}.P_{efg}.P_{sc}$$

where P_{AB} is the probability of a successful Litho-Panspermia as a product of the following probabilities:

P_{biz} = probability that the impactor hits a biologically inhabited zone;
P_{ee} = probability that a rock from the biologically inhabited zone is ejected into an escape orbit;
P_{sl} = probability that an organism ejected with the rock material survives the launch;
P_{ss} = probability that an organism in the ejecta survives in space;
P_{se} = probability that an organism in the ejecta survives entry through the atmosphere of the target planet;
P_{si} = probability that an organism in the ejecta survives the impact onto the surface of the target planet;
P_{rel} = probability that the organism is released from the meteorite;
P_{st} = probability that the environment of the target planet is not toxic to the organism;
P_{sp} = probability that the organism survives potential predators from the target biosphere;
P_{efg} = probability that the environment is favourable for growth and development of the organism;

P_{sc} = probability that the organism and its descendants compete successfully with the indigenous biosphere.

Estimating P_{biz}, and P_{ee}, the probabilities that the impactor hits a biologically inhabited zone and that a rock from that zone is ejected with escape velocity, one has to bear in mind that, at present, the Earth is the only inhabited planet known. According to calculations by C. Mileikowsky and his group, more than 2000 craters of sizes larger than 20 km have been gouged in the surface of the Earth during the last 4 billion years, causing the escape ejection of one trillion fragments from the spall zones of impact craters. Of course, the overwhelming amount of these were cm-sized, probably too small to provide sufficient protection against the harsh environment of interplanetary space. In addition, the spallation model requires that the ejected material is made of hard, dry rocks. It would probably not work with soft soil, mud or poorly consolidated sediments, because such material does not support the spallation process, i.e. the interference of primary and reflected shock waves at a free surface. It has also been argued by B. C. Clark that atmospheric drag might inhibit the escape of very tiny particles, e.g. soil grains, which might even be further dispersed by the impact event.

As seen in Chapter 3, most of the surface of the Earth is teeming with micro-organisms. This life can reach down several kilometres or more. Examples are micro-organisms such as microbial communities living at the surface, or inside rocks, those colonising salt deposits and evaporites (salt crystals formed by evaporation of brines), glacier ice, the vast regions of permafrost, and even sediments down to several kilometres below the ocean floor, as has been described by G. Horneck and C. Baumstark-Khan. The densities of these colonies can be quite high. One gram of soil, for example, may host up to one billion micro-organisms, typically as a mixture of various kinds of vegetative cells which are in a metabolically active state and able to grow and multiply, as well as spores in a dormant state, and which are especially adapted to cope with unfavourable environmental conditions, or unfavourable periods between growing seasons. The fossil record reveals that diverse microbial ecosystems existed on the early Earth, probably as early as 3.5 billion years ago or even earlier. Since then they have continued to thrive, propagate and evolve. Hence, it is very likely, if not inevitable, that rocks ejected from the surface of the Earth after an impact event carry with them billions or even trillions of micro-organisms of different strains, in different developmental stages and of different genealogy.

One of us (G. Horneck), in 2000, has proposed that like the microbial ecosystems mentioned above, similar microbes could be considered on Mars. This is based on the growing evidence that the physical and chemical surface properties of early Earth and early Mars were very similar. If life emerges at a certain stage of planetary evolution when the right environmental, physical and chemical requirements are present, as has been proposed by Nobel laureate Christian de Duve in 1994, then the conditions on early Mars were as favourable for life to emerge as on early Earth. As with the terrestrial microbial communities, several potential protected niches have been postulated for Mars, some of them even for present-day Mars, such as sulphur-rich subsurface areas for chemo-autotrophic communities, rocks for endolithic communities, permafrost regions, hydrothermal vents, soil, or evaporite crystals. Based on the current evidence and on modelling, a deep subsurface liquid water zone is considered as the most favoured potential habitat on Mars. However, this zone may not be located close enough to the Martian surface where the spallation mechanism could work. Anyhow, as long as we are ignorant of whether or not life once started on Mars and, if so, whether and where it might still exist, any estimation on the probability of viable micro-organisms being carried into space with impact ejecta from Mars is highly speculative.

The next factor deals with the nature of the ejection process and P_{sl}, the probability that an organism enclosed in the ejecta would survive the ejection process. The effects of the impacts of km-sized objects on a planetary surface can be studied in ground-based

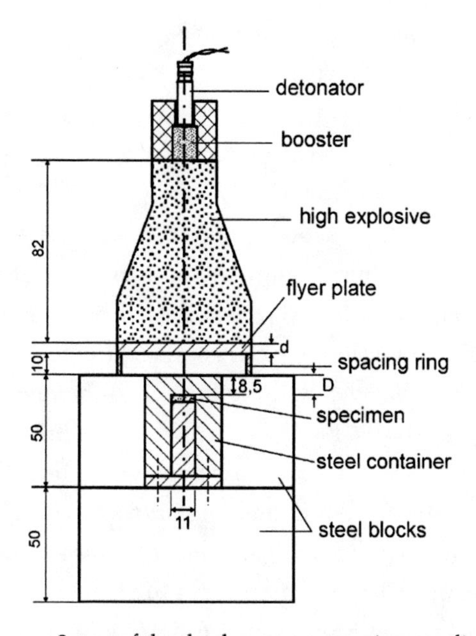

FIGURE 4.5 Set-up of the shock recovery experiments: distances are given in mm, d = thickness of flyer plate, D = thickness of cover plate on specimen (from Horneck and others, 2001).

experiments. Two different experiments have been carried out in this area in order to study the responses of micro-organisms to the stress accompanying such an impact event. These are shock recovery experiments made by G. Horneck and others, and reported in 2001, and acceleration experiments using a gas gun by M. J. Burchell and others, also reported in 2001.

In the shock recovery experiments (Figure 4.5), bacterial spores, the microbial test organisms, were placed between two quartz discs and the sandwich was then shock loaded. The shock wave was reflected several times at the two steel-quartz interfaces corresponding to the medium range of the shock pressures observed in the Martian meteorites. It leads to a shock temperature of 390 °C lasting for a few millionths of a second, and a post-shock temperature of about 250 °C, decaying within minutes. It was found that most of the spores were killed by this shock treatment. However, in each sample, some spores, up to 0.01 %, were able to survive the severe shock pressure and temperature conditions. What do these laboratory experiments tell us? Assuming a terrestrial rock or soil sample inhabited by a hundred million spores per gram – which is a realistic number for the density of soil micro-organisms on Earth – then 1 kg of the rock would accommodate approximately one hundred billion spores. Of these, up to ten million spores could survive even extremely high shock pressures occurring during a meteorite impact. Ballistic experiments have also shown that various kinds of micro-organisms, such as bacterial spores or vegetative cells of the extremely radiation resistant bacterium *Deinococcus radiodurans*, survive gunshots with accelerations up to 4500 km/s ($460\,000 \times g$). In similar experiments, R. Mastrapa and others demonstrated a survival rate of nearly 90 % for the spores of the bacterium *Bacillus subtilis*. Hence, accelerations and shocks as they occur during planetary ejections are apparently not a barrier to the interplanetary transfer of life.

Once rocks have been ejected from the surface of their home planet, the microbial passengers now face an entirely new set of problems. These are related to exposure to the space environment. The probability for micro-organisms in the ejecta to survive in space is described by P_{ss}. In order to study the survival of resistant microbial forms in the upper atmosphere or in free space, microbial samples have been exposed *in situ* by the use of balloons, rockets and spacecraft, and their responses investigated after recovery.

All this work has been reviewed by two of us (G. Horneck and A. Brack) in 1992 and again by one of us (G. Horneck) in 1993.

For this purpose, several facilities have been developed, such as an exposure device on the Gemini missions, the MEED experiment on Apollo, the ES029 experiment on the Spacelab 1 mission, the Exobiology and Radiation Assembly (ERA) on the EURECA free-flying platform, the UV-RAD experiment on the Spacelab D2 mission and the BIOPAN facility for the FOTON Russian recoverable capsule.

These investigations were supported by studies in the laboratory in which certain parameters of space (high and ultrahigh vacuum, extreme temperatures, UV radiation of different wavelengths and

FIGURE 4.6 The ESA EURECA free-flying platform being deployed by the Space Shuttle. (Courtesy NASA.)

ionising radiation) were simulated. In these experiments the microbial responses (physiological, genetic and biochemical changes) to the various exposure factors, either applied separately or in combination, have been determined.

In order to understand the physical stresses imposed on spores in space, it is first necessary to understand the space environment itself. The environment in Earth orbit, where most of the experiments were performed, is characterised by a high vacuum, an intense radiation climate, and extremes of temperature (Table 3).

In free interplanetary space, very low pressures prevail and the vacuum is very high. In the vicinity of a planetary or other body, the pressure may significantly increase due to outgassing. In a low Earth orbit, the major constituents of the environment are molecular oxygen and nitrogen, as well as highly reactive oxygen and nitrogen atoms. In the vicinity of a spacecraft, the pressure further increases, depending on the degree of outgassing.

The radiation environment of our Solar System consists of rays of galactic and of solar origin (Figure 4.8). The galactic cosmic radiation

TABLE 3 The physical conditions in interplanetary space and in low Earth orbit.

Space parameter	Interplanetary space	Low Earth orbit ($\leq$500 km)
Space vacuum		
Pressure (Pa)*	10^{-14}	10^{-6}–10^{-4} [a]
Residual gas (part/cm^{-3})	1 H	1×10^5 H 2×10^6 He 1×10^5 N 3×10^7 O spacecraft atmosphere [b]
Solar electromagnetic radiation		
Irradiance (W/m^2)	Different values [c]	1380
Spectral range (nm)	Continuum	Continuum
Cosmic ionising radiation		
Dose (Gy/a)	$\leq$0.1 [d]	400–10 000 [e]
Temperature (K)	>4 [c]	wide range [c]
Microgravity (g)	$<10^{-6}$	10^{-3}–10^{-6}

*1 Pa = 10^{-5} bar

[a] Depending on outgassing of the spacecraft

[b] Sources of contamination: waste dumping (H_2O, organics); thruster firing (H_2O, N_2O, NO)

[c] Depending on orientation and distance to Sun

[d] Depending on shielding, highest values at mass shielding of 0.15 g/cm^2

[e] Depending on altitude, orbit and shielding, highest values at high altitudes and mass shielding of 0.15 g/cm^2

entering our Solar System is composed of protons, electrons, α-particles (He atoms) and heavy ions of charge $Z > 2$, the so-called HZE particles (particles of *h*igh charge *Z* and high energy *E*). The solar particle radiation, carried by the solar wind and during solar flares (solar particle

FIGURE 4.7 A laboratory facility simulating space conditions and used to study the effects of these on micro-organisms. (Courtesy DLR.)

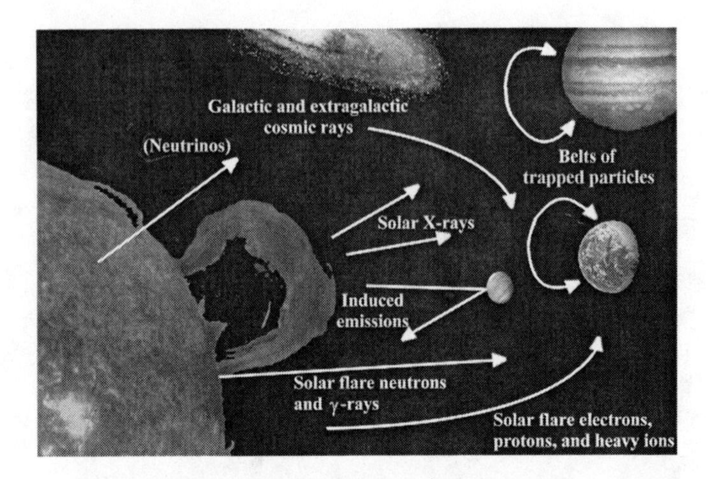

FIGURE 4.8 Space radiation sources in the Solar System (from Baumstark-Kahn and Facius, 2002).

events), is composed of protons, α-particles and a relatively small number of heavier ions. In the vicinity of the Earth, in the radiation belts, protons and electrons are trapped by the geomagnetic field, thereby protecting the Earth from their effects.

The spectrum of solar electromagnetic radiation spans several orders of magnitude, from short wavelength X-rays to long-wavelength radio frequencies. At the distance of the Earth (1 astronomical unit or AU), the solar irradiance amounts to 1360 watts per square metre. Of this radiation, 45 % is in the infrared part, 48 % in the visible and only 7 % in the ultraviolet range. The extraterrestrial solar spectral UV irradiance has been measured during several space missions, such as the Spacelab 1 mission and the EURECA mission in 1992–1993 (Figure 4.6).

The temperature of a body in space depends on its position with respect to the Sun and other orbiting bodies, its surface, size, mass and its reflectivity or albedo. In Earth orbit, the energy sources include solar radiation, the Earth albedo and terrestrial radiation. Periodically, an orbiting object can be both directly exposed to and then shaded from the Sun as it passes successively through the Earth's day and night sides. So, in Earth orbit, a body can experience very large temperature swings.

How can terrestrial micro-organisms cope with these harsh conditions in outer space? If our terrestrial organisms originated just on Earth then there was probably never a need for them to evolve and adapt to conditions comparable to those in interplanetary space. Therefore, perfectly suitable test systems adapted to the hostile conditions of space are not to hand. On the other hand, a variety of organisms do exist that are adapted for growth and survival in some of the extreme conditions of our biosphere. Some of these may be suitable candidates for studies of micro-organisms in space. The rationale behind such studies is that if terrestrial organisms can survive the rigours of space, it is more likely that interplanetary transfer is a feasible process.

Several procaryotic and eucaryotic micro-organisms can survive unfavourable conditions in a dormant state, and are capable of full metabolic recovery when conditions become more favourable again. Bacterial examples are the *Bacillus* and *Clostridium* species, which are capable of producing dormant spores. In these spores, the DNA is extremely well protected against environmental stress, such as desiccation, oxidising agents, UV and ionising radiation, low and high pH, as well as temperature extremes. In 2000, W. L. Nicholson and others reported that the high resistance of *Bacillus* spores is mainly due to a dehydrated core enclosed in a thick protective envelope, the cortex and the spore coat layers (Figure 4.9), and to the saturation of their DNA with small, acid-soluble proteins whose binding greatly alters the chemical and enzymatic reactivity of the DNA. These *Bacillus* spores are the champions of longevity. They can survive in the dormant state for thousands of years. The isolation of viable *Bacillus* spores from Dominican amber with ages of several millions of years, and the isolation of a halotolerant spore-forming bacterium from within a brine inclusion inside a halite crystal from a 250 million-year-old salt deposit have all been demonstrated. This has all been reviewed by W. L. Nicholson in 2003. Due to their exceptional resistance to environmental extremes and their extreme longevity, spores of *B. subtilis* have been subjected to intense tests on their survival in space. This in turn has been reviewed by G. Horneck and others in 2002.

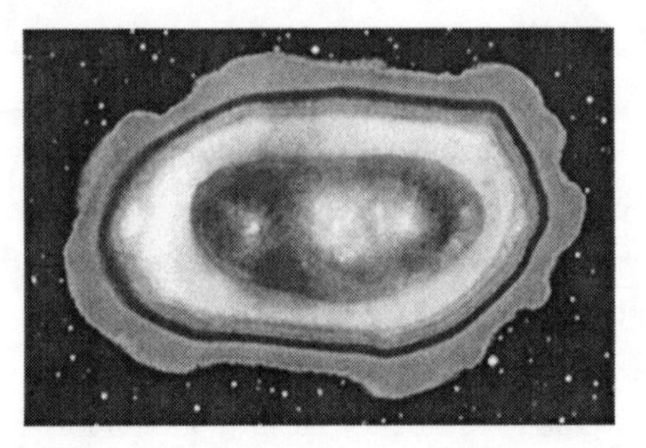

FIGURE 4.9 *Bacillus subtilis* spore, the champion of longevity. (Courtesy of S. Pankratz.)

However, since the *Bacillus* species are not found at the deepest branch of the 'Tree of Life' (see Chapter 3), they are certainly not considered representative of a hypothetical microbial 'first coloniser' on the early Earth. They have, however, been mainly used as model systems for many of the studies in space.

Since this scenario calls for solid rocks that are expelled from planetary surfaces during impact events, microbial communities living inside rocks, i.e. endolithic micro-organisms, might be more suitable candidates than bacterial spores for such space odysseys. Endolithic life forms, such as cyanobacteria, algae, fungi and lichens, can all survive in extreme habitats. Examples have been found in rocks of the dry valleys of Antarctica as well as in hot deserts. E.I. Friedmann has shown that within the rock there is a favourable microclimate as opposed to the hostile macroclimate of low humidity, temperature extremes, and a high bombardment of solar radiation, all outside the rock (see Chapter 5). In addition, cyanobacteria have developed a variety of Sun-protective mechanisms, such as the accumulation of UV-B absorbing pigments and photo-repair mechanisms for DNA and other functional molecules. Most lichens are extremely tolerant to desiccation. They occur in the driest and coldest regions on Earth. Endolithic microbial communities will be studied in space on the platform EXPOSE (Table 4) that will be mounted on the

TABLE 4 Experiments to be carried out onboard EXPOSE.

- Interstellar organic chemistry
- Photochemical processing, emergence of prebiotic molecules
- Role of ozone layer in protecting the biosphere against solar UV-B
- Protection of endolithic and osmophilic organisms in space
- Protection of spores in artificial meteorites
- DNA photoproducts from extraterrestrial UV radiation
- Photochemistry of polycrystalline uracil
- Mutational spectra in spores produced by vacuum and solar UV

external platform of the European Columbus Laboratory of the International Space Station (Figure 4.10).

Other interesting cases are micro-organisms in salts, called evaporites. L. J. Rothschild and others have reported bacteria, cyanobacteria and algae trapped in evaporite deposits that show metabolic activity, although at a very slow rate. These micro-organisms are capable of fixing carbon and nitrogen, which allows them to survive while trapped in salt and even to maintain all the functions necessary for active life (Plate VII). Endoevaporitic micro-organisms were first exposed to the space environment on ESA's exposure facility BIOPAN (Figure 4.13), and they will be further studied on the EXPOSE platform (Table 4). This will help to determine the role of endogenous protection (e.g. by the carotenoid pigment) and exogenous protection (e.g. by salt crystals) against the deleterious effects of space, particularly solar UV radiation and vacuum.

Cosmic radiation is a deadly aspect of the conditions in outer space. Hence, extremely radiation-resistant micro-organisms, such as *Deinococcus radiodurans*, which are the most radiation-resistant bacteria known to exist, have been used for studies in space. V. Mattimore and J. R. Battista have reported that although *D. radiodurans* is non-sporulating, it can become dormant under certain environmentally adverse conditions, such as lack of food, desiccation or

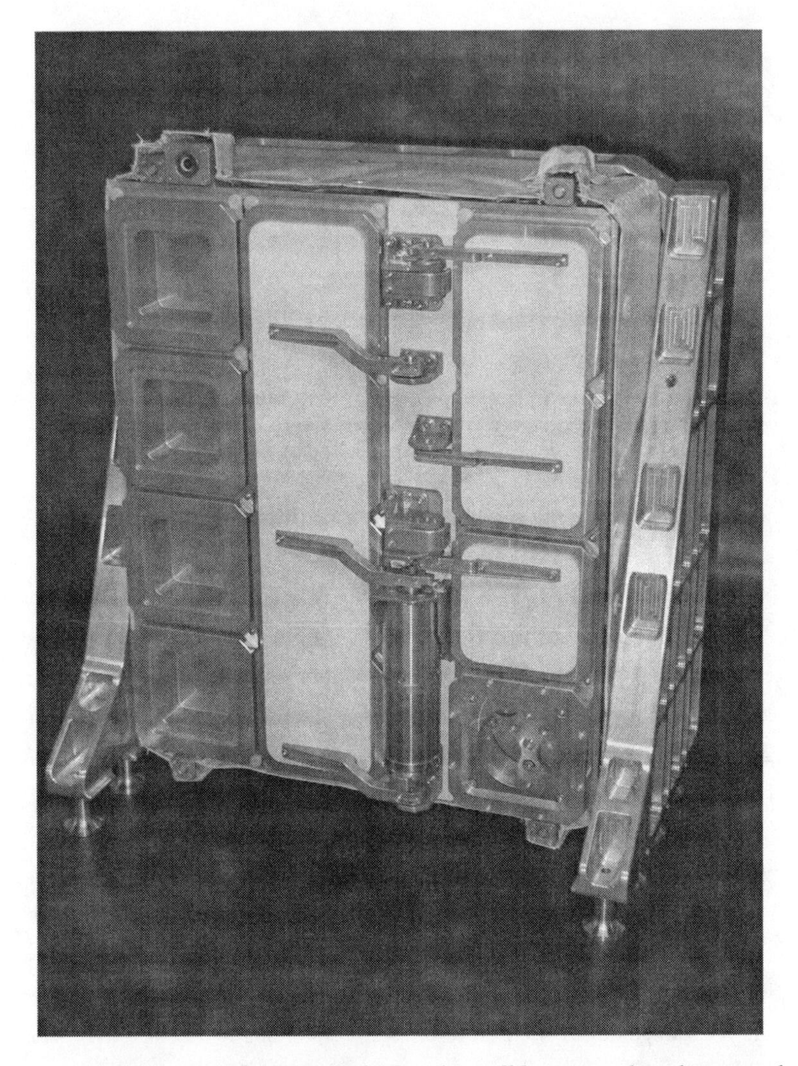

FIGURE 4.10 The EXPOSE facility that will be mounted on the external part of the Columbus Laboratory on the ISS. (Courtesy ESA.)

low temperatures. During dormancy, its metabolism slows down. As a consequence, the radiation damage to its DNA accumulates. However, *D. radiodurans* possesses a very efficient DNA repair mechanism which restarts when dormancy ends. K. Dose and others exposed this very common soil micro-organism to space during the European EURECA mission in 1992–1993.

Although most studies in space or under simulated space conditions have been performed on procaryotic micro-organisms and their cellular biomolecules, higher eucaryotic organisms that survive complete dehydration might also be suitable test systems. These organisms include the slime mould *Dictyostelium*, dry active baker's yeast, brine shrimp cysts of *Artemia salina*, dry larvae and the adults of several species of nematodes. In some of these organisms, surviving dehydration is related to the accumulation of substances called polyols, in particular, trehalose.

In parallel to experiments in space, ground control experiments have been carried out, in many cases at the space simulation facilities of the German Aerospace centre DLR (Figure 4.7). The major results obtained from space experiments over more than 20 years can be summarised in the following paragraphs.

Because of its extreme dehydrating effect, the space vacuum has been argued by M.D. Nussinov and S.V. Lysenko to be one of the factors preventing interplanetary transfer of life. Indeed, most organisms are killed when exposed to vacuum conditions. If not protected by internal or external substances, cells in a vacuum experience dramatic changes in such important biomolecules as lipids, carbohydrates, proteins and nucleic acids. Upon desiccation, the lipid membranes of cells undergo dramatic phase changes from planar bilayers to cylindrical bilayers (Figure 4.11). The carbohydrates, proteins and nucleic acids give products that become crosslinked, eventually leading to irreversible polymerisation of the biomolecules. Along with these structural changes are functional changes, including altered selective membrane permeability, inhibited or altered enzyme activity, decreased energy production and the alteration of genetic information, etc.

However, G. Horneck has reported in 1993 that space experiments have shown that up to 70 % of bacterial and fungal spores can survive a short-term (for example, 10 days) exposure to space vacuum, even without any protection – as long as they are shielded against the flux of the extraterrestrial solar UV rays. The chances of survival in space vacuum are even increased if the spores are embedded in

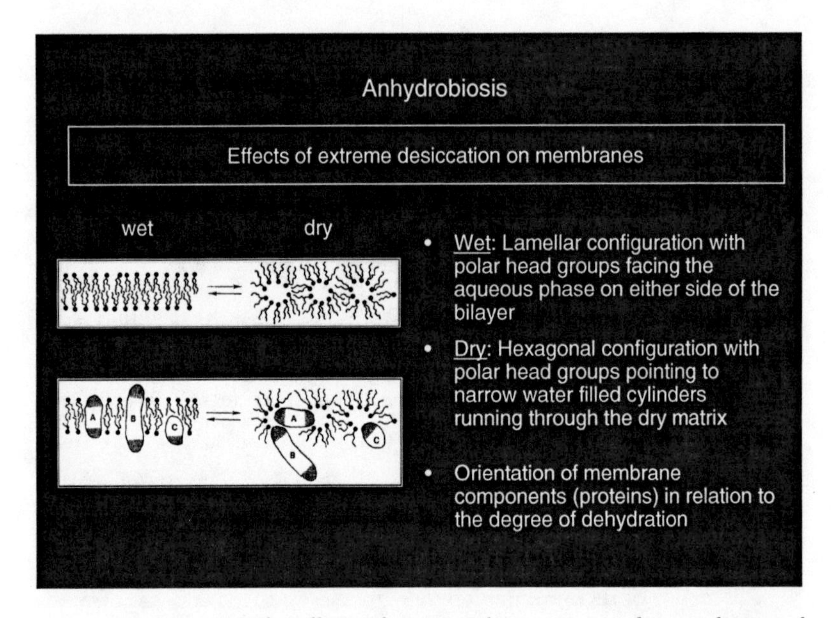

FIGURE 4.11 The effects of extreme desiccation on the membranes of biological cells. (Courtesy DLR.)

protecting substances, such as sugars, or salt crystals, or if they are exposed in thick layers. For example, during the NASA mission of the *L*ong *D*uration *E*xposure *F*acility, (LDEF see Figure 4.12), in the German experiment Exostack, 30 % of *B. subtilis* spores survived nearly six years of exposure to space vacuum when embedded in salt crystals, whereas 80 % survived in the presence of glucose. R. Mancinelli and others have shown that crystalline salt also provided sufficient protection for osmophilic microbes to survive at least two weeks in the vacuum of space. Sugars and polyalcohols stabilise the structure of the cellular macromolecules, especially during vacuum-induced dehydration, leading to increased rates of survival. This is because they help to prevent damage to the DNA, membranes and proteins by replacing the water molecules during the desiccation process and thereby preserving the three-dimensional structure of the biomolecules. Endolithic (that is, those lodged in stones) micro-organisms are embedded in a complex extracellular polysaccharide matrix which might also be expected to provide a degree of protection

FIGURE 4.12 The LDEF mission, where spores of *B. subtilis* were exposed to space conditions for more than six years. (Courtesy NASA.)

from extreme vacuum, such as might happen after a sample is blasted into space during an impact event.

The high resistance of bacterial endospores to desiccation is mainly due to the dehydrated strongly mineralised core surrounded by several protective layers (Figure 4.11) and the stabilisation of their DNA by specific proteins. The space experiments have shown that the addition of glucose to the spores further increased their survival rate in vacuum.

Although spores of *B. subtilis* survived vacuum treatment perfectly well, a tenfold-increased mutation rate over the spontaneous rate has been observed after exposure to space vacuum. Specific molecular changes in the DNA, as well as DNA strand breaks, are conjectured to be responsible for the mutations induced in spores in vacuum. Because DNA repair is not active during the dormant state of a bacterial spore, such damage to the genetic material will accumulate during long periods in space. Hence, the genetic integrity of bacterial spores after exposure to space ultimately depends on the efficiency of DNA repair processes that only resume after rehydration and germination of the spores.

The next harmful parameter of space to be considered is the solar UV radiation. In particular, the highly energetic, short-wavelength part of the solar spectrum is of high biological effectiveness, because it is directly absorbed by the DNA molecules of the cells, causing photochemical changes of the DNA bases. If this damage to the DNA is not repaired, the cells will mutate or ultimately die. Thanks to the effective shielding by the stratospheric ozone layer, our biosphere is safely protected from this highly damaging portion of the UV rays. However, this was not always so. For the first 2 billion years of life's existence on Earth, the stratospheric ozone and its UV screening effect had not yet been built up. Hence, early life was probably forced to stay away from the intense sunshine, and to hide deep within the oceans, in subsurface regions or in caves.

Because of its direct damage to the DNA of cells, the solar UV radiation has been found to be the most dangerous element of the space environment, as has been shown by tests on dried preparations of viruses, and of bacterial and fungal spores. G. Horneck and A. Brack reported in 1992 that the full spectrum of extraterrestrial UV radiation has been shown to kill unprotected spores of *B. subtilis* within seconds. This high biological effectiveness of the extraterrestrial UV radiation is mainly caused by damage to the spore's DNA induced by the UV-C part of the spectrum, as identified by spectroscopy. Compared to the terrestrial UV radiation climate, the extraterrestrial UV spectrum is about one thousandfold more effective in killing the spores.

In order to investigate the combined action of solar UV radiation and vacuum, a series of experiments were carried out using the BIOPAN facility of ESA onboard the Russian FOTON recoverable satellite (Figure 4.13). *B. subtilis* spores were exposed to space, either without any protecting agent, or mixed with clay, red sandstone,

FIGURE 4.13 The ESA BIOPAN facility tests the effects of extraterrestrial solar UV radiation and space vacuum on the survival of bacterial spores and osmophilic micro-organisms. (Courtesy ESA.)

simulated Martian soil or meteorite powder, in dry layers as well as in so-called 'artificial meteorites', i.e. centimetre cubes filled with clay and spores in naturally occurring concentrations. This has all been reported by G. Horneck and others in 2001. After about two weeks in space, it was found that unprotected spores in layers either open to space or behind a quartz window were completely or nearly completely inactivated (only one spore in a million or less survived). The same low survival was obtained behind a thin layer of clay acting as an optical filter. However, the survival rate was increased by 100 000 and more if the spores in the dry layer were directly mixed with powdered clay, rock or meteorites, and an up to 100 % survival rate was reached in the 'artificial meteorites', i.e. soil mixtures with spores comparable to the natural soil-to-spore ratio. These data confirm the deleterious effects of extraterrestrial solar UV radiation. Thin layers of clay, rock or meteorite were only successful in UV-screening if they were in direct contact with the spores. These data suggest that in the scenario of Litho-Panspermia, small rock ejecta a few centimetres in diameter could be sufficiently large to protect bacterial spores against the effects of the intense UV radiation in space. However, micron-sized grains – that is, single spores – as originally envisaged by classical Panspermia, are certainly not sufficiently protected to survive space conditions.

During the same series of experiments R. Mancinelli and others, using the BIOPAN facility, showed that crystalline salt provides a certain protection for microbes in space. For example, a species of the cyanobacterium *Synechococcus* that inhabits gypsum-halite crystals was capable of nitrogen and carbon fixation, and about 5 % of a species of the extreme halophile *Haloarcula* survived after exposure to the space environment for two weeks.

The question remains whether a layer of meteorite material a few centimetres thick would also provide sufficient shielding against the damaging attacks of the ionising components of radiation in space, especially in view of the extended periods in space, as foreseen in the Litho-Panspermia scenario. Although the flux of the heavy primaries of galactic cosmic radiation, the so-called HZE particles, is very low (they contribute to only about 1 % of the flux of particulate radiation in space), they are considered as the biologically most aggressive species of cosmic radiation. Such HZE particles of cosmic radiation are conjectured to set the ultimate limit on the long-term survival of spores in space because they penetrate even thick shielding.

Because of the low flux of these HZE particles, methods have been developed to localise precisely the trajectory of individual HZE particles relative to a biological object, and to correlate the physical data of the particle relative to the observed biological effects along its path. In the Biostack method, visual track detectors were sandwiched between layers of biological objects in a resting state, e.g. *B. subtilis* spores. This method allows (i) localisation of each HZE particle's trajectory in relation to the biological specimens; (ii) investigation of the responses of each biological individual hit separately as caused by the radiation; (iii) measurement of the impact parameter b (i.e. the distance between the particle track and the sensitive target); (iv) determination of the physical parameters [charge (Z), energy (E) and linear energy transfer (LET)]; and finally (v) to correlate the biological effects with each HZE particle parameters, as has all been described by G. Horneck in 1992.

Among other biological specimens, *B. subtilis* spores were used as test systems. It was found that a certain fraction of the spores (about 30 %) survived even a central hit of an HZE particle of cosmic radiation. Taking these data and the low flux of the HZE particles in space, it has been calculated that it may take up to several hundred thousand to one million years for a single spore in space to be killed by an HZE particle.

C. Mileikowsky and his group arrived at comparable timescales. They base their calculations on the assumption that the rock may accommodate about one hundred million spores per gram (which is a realistic number for terrestrial conditions), of which at least 100 spores per gram would survive, i.e. one survivor per million spores. This survival rate, which is still sufficiently high for Litho-Panspermia to be successful, would be reached after about 1 million years for spores located in the rock behind 1 m of shielding, after about 300 000 years behind 10 centimetres of shielding (maximum dose rate because of secondary radiation produced by the interaction of cosmic rays in the first 10–20 cm of the shielding material) and after about 600 000 years without any shielding. In larger rocks, e.g. providing 2 to 3 metres of meteorite shielding material, the spores may travel for up to 25 million years in space for the same survival rate. B. C. Clark arrived at a 'survival time' of 1.5 million years behind a meteorite shielding of 10 centimetres and 7.5 million years at a depth of 1 metre.

The data obtained so far on the responses of resistant micro-organisms to the complex environment of space support the conclusion that space – although it is very hostile to terrestrial life – is not a barrier for cross-fertilisation of planets within the Solar System. This has been concluded from all the experiments in space described above.

For future research on bacterial spores and other micro-organisms in space, ESA has developed the EXPOSE facility that is now planned to be attached to the external platform of the Columbus module of the International Space Station (ISS) for 1.5 years (Figure. 4.10). EXPOSE will support long-term *in-situ* studies of microbes in artificial meteorites, as well as of microbial communities from special ecological niches, such as endolithic and endoevaporitic ecosystems (Table 4). These experiments include the study of photobiological processes in simulated radiation climates of planets (e.g. early Earth, early and present-day Mars, and the role of the ozone layer in protecting the biosphere from harmful UV-B radiation), as well as studies of the probabilities and limitations for life to be successfully exported beyond its planet of origin.

If spores or other micro-organisms trapped inside ejecta survive the impact and then the transit through space, they must be captured by the gravity of a recipient planet and survive entry. This is covered by the terms P_{se}, the probability that an organism in the ejecta survives entry through the atmosphere of the target planet, and P_{si}, the probability that an organism in the ejecta survives the impact with the surface of the target planet.

When captured by a planet with an atmosphere, most meteorites are subjected to very high temperatures during entry. The fate of the meteorite strongly depends on its size. Small fragments (e.g. 1 g – 1 kg) may burn up entirely in the Earth's atmosphere, whereas larger fragments are slowed dramatically upon entry to a terminal velocity of a few hundred metres per second. Micrometeorites of a few μm in size may even tumble through the atmosphere without being heated above 100 °C. Very large meteorites may break into pieces upon entering the atmosphere. During entry, medium sized meteorites and the fragments of larger ones may develop a melted crust, whereas the inner part still remains cool. Because the entire entry process takes only a few seconds, the outermost layers form a kind of heat shield and the heat does not reach the inner parts of the meteorite. Hence, the interior of the meteorite is not heated significantly above its in-space temperature, except for a few millimetres or centimetres of ablation crust at the surface. Upon impact, the meteorite or its fragments are further shattered and mixed with the surface material of the planet. It is therefore possible not only for endolithic microbes to survive entry

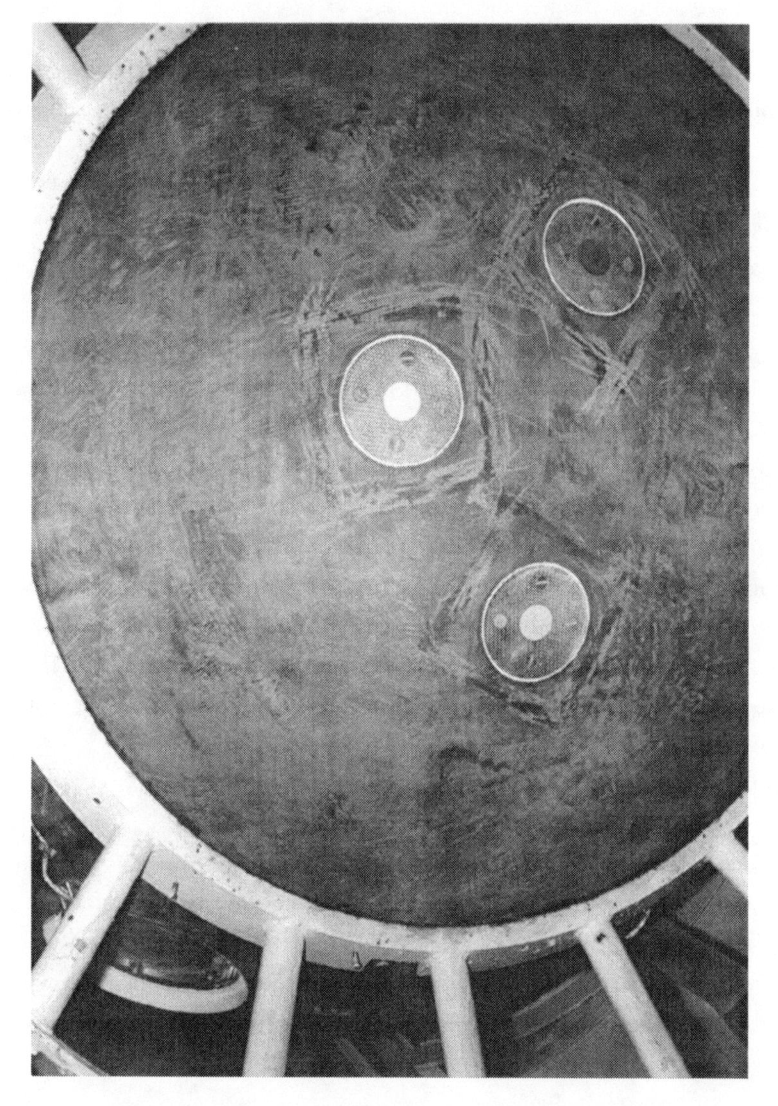

FIGURE 4.14 The STONE experiment on the FOTON recoverable satellite. (Courtesy ESA.)

embedded in a meteorite, but actually be injected into the recipient planet's crust, thereby encountering an environment potentially conducive to growth. This all allows some estimate to be made about P_{rel}, the probability that the organism is released from the meteorite.

Although the data obtained from meteorites suggest that a substantial number of microbes may survive the landing process on a planet, so far, no experiments have been done to investigate the effects of the landing process experimentally. Recently, ESA has developed a facility, called STONE, which is attached to the heat shield of a FOTON satellite to test mineral degradations during landing (Figure 4.14). This facility will be an ideal tool for studying the effects of landing on bacterial spores embedded in an artificial meteorite.

In order to assess the chances of survival and proliferation of micro-organisms on the recipient planet, covered by the probabilities P_{st} (the probability that the environment of the target planet is not toxic to the organism), P_{sp} (the probability that the organism survives potential predators from the target biosphere), P_{efg} (the probability that the environment is favourable for growth and development of the organism), and P_{sc} (the probability that the organism and its descendants compete successfully with the indigenous biosphere), experiments with terrestrial micro-organisms or selected communities from extreme environments under conditions simulating those of other planets, e.g. Mars, will be required.

Therefore, based on recent observations, experimental results and calculations, we can conclude that all the steps required for a viable transport of micro-organisms between the planets of a solar system can be fulfilled by the Litho-Panspermia mechanism. This covers the escape process, the interim state in space, i.e. survival of the biological material over timescales comparable with the interplanetary passage, and the entry process, i.e. non-destructive deposition of the biological material on another planet.

5 What are the signatures of life?

Life on or near the surface of a planet will almost certainly interact with the soil, rocks, atmosphere and oceans with which it is in contact. Since life creates order in its environment, it acts as a driving force for a number of global chemical transformations. On Earth, typical examples of such life-induced chemical anomalies are the excess of atmospheric oxygen and the release of large quantities of hydrogen sulphide by sulphate-reducing bacteria in the oceans. Also, the persistence of metastable gas mixtures such as O_2, N_2 and CH_4 in the atmosphere, and of isotopic disequilibria (e.g. between water-bound oxygen of the hydrosphere and atmospheric O_2), are driven by the effects of the biosphere on its environment.

Conspicuous anomalies within the atmosphere and oceans of a planet may, therefore, be taken as a-priori evidence of the presence of life, as proposed by British scientist James Lovelock in the 1960s. Applying these ideas to the Earth's biosphere led Lovelock to his Gaia hypothesis, in which the biosphere as a whole is considered to be a living entity. Applying this criterion to the present composition of the Martian atmosphere led American scientist Tobias Owen and others to conclude that there is little, if any, indication of contemporary biological activity on Mars.

In addition, a planetary biosphere can leave discrete vestiges in the surrounding rocks. On Earth, organisms commonly leave a morphological and biochemical record of their former existence in sedimentary rocks. Though in part highly selective, this record may survive, under favourable circumstances, over billions of years before being destroyed during any metamorphic changes of the host rock. This is particularly true for relics of multicellular life, but also – albeit with restrictions – for micro-organisms that dominated the Earth during the first 3 billion years of geological history.

EVIDENCE OF EXTANT LIFE

The microbial world

Microbial procaryotes have flourished on Earth for more than 3.5 billion years. They dominated the Earth's biosphere during the first 2 billion years before the first unicellular mitotic eucaryotes appeared. Therefore, in a search for extant life beyond the Earth, micro-organisms are the most likely candidates for living things. Structural, functional and chemical characteristics have been frequently used to identify terrestrial microbial communities. This includes signatures for actively metabolising or even proliferating life forms as well as for resting life forms, such as dehydrated or frozen organisms or bacterial spores. Special methods are used to detect microbes in extreme environments, such as hot vents, permafrost, permanent ice, subsurface regions, high atmosphere, rocks and salt crystals. These environmental extremes may be considered as terrestrial analogues of possible ecological niches on Mars and on other selected planets and moons of our Solar System.

Structural indications of life

Cells and subcellular structures can be detected directly by microscopy. Microbial cells have been observed in samples collected from natural environments, such as smears of soil or subsurface aquifer sediments, droplets taken from hydrothermal vents or brine, airborne micro-organisms trapped on a sticky surface, or thin fractures or slices of rocks or salt crystals. Most of them are ovoid, spherical or rod-shaped objects of a few microns in size. The observation of dividing cells, i.e. paired cells or cells in chains, indicate that the cells were actively growing *in situ* immediately before or during the sampling procedure; this is direct evidence of how microbes interact with their environment. Details on specific means to differentiate between cellular

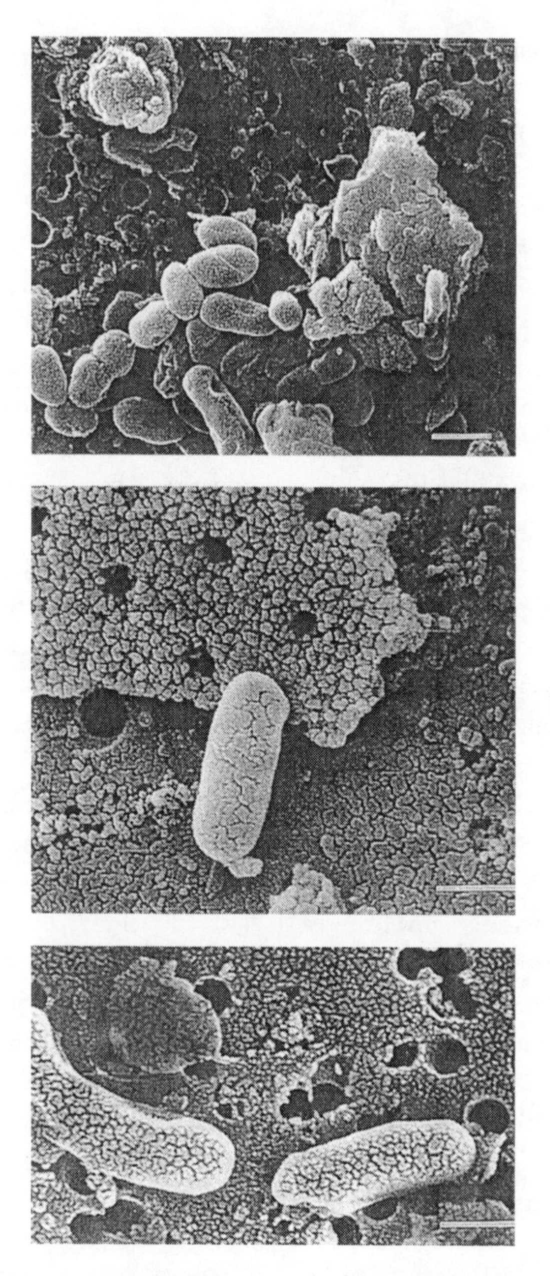

FIGURE 5.1 Micro-organisms in ice-core samples from ancient layers of the Antarctic ice sheet at depths of 1.6 m (top) and 2.4 m (middle and bottom). The bar represents 1 μm. (Scanning electron micrograph images courtesy of S. S. Abyzov.)

structures and similar structures of non-biological origin are given in Appendix 5.

Large microbial communities are also macroscopically visible. Examples have been found by Rothschild and others in evaporites of up to 4 cm in thickness which contain one or two horizontal coloured bands several mm in thickness well below the surface, as shown in Plate VII (see also Chapter 4). These layers appear in different colours, from tan to pink-brown to green or green-grey within a basically white salt matrix. E. I. Friedmann and J. A. Nienow found similar signatures of microbial communities colonising sandstone in cold and hot deserts.

The colonised zone below the rock crust reaches up to 10 mm deep and is composed of distinct parallel bands of black, white, green and blue-green. Throughout the colonised zone, the quartz is rendered colourless, probably by oxalic acid secreted by the organisms themselves, and the intense sunlight of the desert is strongly attenuated by the overlying mineral layers. Hence, these microbial communities have conquered an ecological niche that allows micro-organisms to live in an otherwise hostile environment.

S. E. Campbell has described desert crusts, formed by microorganisms, which may cover even larger areas of hundreds of square metres with a thickness of a few centimetres. These mats, dominated by cyanobacteria, show features such as sediment trapping and accretion, a convoluted surface and polygonal cracking. Sand and clay particles are trapped within a network of filamentous cyanobacteria which secrete mucous sheets to which the particles adhere. J. F. Stolz and Lynn Margulis have described comparable large areas of stratified mats of microbial communities observed in the salt flats of high-salt lagoons, such as the Laguna Figueroa on the Pacific coast of Baja California and also in a host of related habitats.

Evidence of microbial activity as a functional characteristic of life

The most unequivocal indication of microbial life is obtained by the isolation of micro-organisms in pure culture from a sample under investigation, followed by a detailed analysis for their biochemical properties. Microbial isolates have been obtained from diverse extreme habitats. Examples are deep crystalline rock aquifers several hundreds of metres below the surface, as studied by T. O. Stevens and J. P. McKinley, the interiors of ice-cores from drillings in the Antarctic ice down to a depth of several hundred metres, as carried out by S. S. Abyzov, and cores from drillings in permafrost regions in Siberia done at similar depths by D. A. Gilichinsky and others. As was seen above, it was found that the interiors of rocks in cold and hot deserts provide ecological niches for microbial communities, described by J. Siebert and P. Hirsch, just as do crystalline salts from evaporite deposits.

Micro-organisms have been isolated by H. S. Vishniac from extremely cold environments, such as Antarctic soils. They have also been isolated from hot environments at temperatures of 80–115 °C, which are usually associated with active volcanism as hot springs, solfataric fields, shallow submarine hydrothermal vents, abyssal hot vent systems (black smokers), as well as oil-bearing deep geothermally heated soils, described by Karl Stetter. Microbial communities have been found buried in river and lake sediments by J. Kolbel-Boekel, and in the upper regions of the atmosphere up to an altitude of about 70 km by A. A. Imshenetsky and others.

In 1995, R. J. Cano and M. K. Borucki reported the successful isolation and revival of bacterial spores from the abdomens of 25–40-million-year-old bees that had been preserved in amber. Details on the use of nucleic acid sequencing technology in this area are given in Appendix 5.

What about the search for extant life carried out by the US Viking missions at the surface of Mars over 25 years ago? The life-detection instrument package of the Viking missions was comprised of three experiments to detect any metabolic activity of potential microbial soil communities:

1. The pyrolitic release experiment tested carbon assimilation, i.e. photoautotrophy, as a method of incorporating radioactively labelled

FIGURE 5.2 A view from the Viking spacecraft sitting on the Martian surface. (Courtesy NASA.)

carbon dioxide in the presence of sunlight. This has been discussed by N. H. Horowitz and others.

2. The labelled release experiment tested catabolic activity, e.g. digestion of food, as a metabolic capability to release radioactively labelled carbon from organic nutrient compounds, as described by G. V. Levin and P. A. Straat.
3. The gas-exchange experiment tested respiration as a metabolic production of gaseous by-products in the presence of water and nutrients, as described by V. I. Oyama and others.

At first, all three Viking biology experiments appeared to give results indicative of active chemical or even biological processes. However, gas chromatography/mass spectrometry (GCMS) revealed a complete lack of any organic residues in the Martian soil and non-biological processes are now assumed to be responsible for the observed reactions. This has been discussed by H. P. Klein. So far, the mechanisms underlying the reactions of the biology experiments are not known. The most likely explanation is that highly reactive peroxides are produced in the soil by the intense unfiltered solar UV radiation, which would also account for the lack of organics in the samples, as reviewed by G. Horneck. The conclusion is that extant life is absent at the two Viking sites.

Chemical signatures and biomarkers

Micro-organisms that are isolated and enriched from their primary habitats can be subjected to a thorough biochemical analysis, such as determination of the total biomass and the content of proteins, nucleic acids, lipids, sugars etc. The detailed characterisation of the pattern of biomarker mixtures from a sample permits the assessment of the major contributing species and their metabolic state, e.g. actively metabolising, dormant or extinct. F. Chapelle has explained how phospholipids, which are part of every bacterial membrane, are valuable indicators of the living biomass that is present because they maintain a relatively constant portion of the cell mass and disintegrate fairly rapidly after cellular death. The pattern of lipids is also diagnostic for different species of a microbial community, such as phototrophic cyanobacteria, photosynthetic bacteria, bacterial heterotrophs and archaea, as discussed by R. E. Summons and others.

The oldest sedimentary rocks on Earth have been found in Greenland. Analysis of the organic carbon in these sediments indicates that life may be more than 3.7 billion years old. This has been reported by M. Schidlowski in 1988, S. J. Mojsis and others in 1996, P. W. U. Appell and S. Moorbath in 1999, and M. T. Rosing in 1999. This isotopic evidence stems from the fact that the carbon atom has two stable isotopes, carbon 12 and carbon 13. The $^{12}C/^{13}C$ ratio in non-biological mineral compounds is 89. In biological material, the process of photosynthesis gives a preference to the lighter carbon-12 isotope and raises the ratio to about 92. Consequently, the carbon residues of previously living matter may be identified by this enrichment in ^{12}C. A compilation has been made of the carbon isotopic composition of over 1600 samples of fossil kerogen (a complex organic macromolecule produced from the debris of biological matter) in these sediments and compared with that from carbonates in the same sedimentary rocks. This showed that the biosynthesis by photosynthetic organisms was involved in all the sediments studied. In fact, this difference is now taken to be one of the most powerful indications that life on Earth was active nearly 3.9 billion years ago because the samples include specimens from across the Earth's geological timescale. This conclusion is consistent with the diversity of the 3.5 billion-year-old fossilised microflora found by Frances Westall and others in the Barberton and Pilbara greenstone belts in South Africa and Australia, respectively. Since the age of the Earth is 4.5 billion years, this means that life began as a quite early event in the Earth's history. Unfortunately, the direct clues that could help us identify the molecules which participated in the emergence of life on Earth about 4 billion years ago have been erased by the combined action of plate tectonics, the permanent presence of running water, the unshielded solar ultraviolet radiation and atmospheric oxygen.

Details on the use of optical handedness measurements to detect extant life are given in Appendix 5.

A detailed chemical analysis of the surface soil of Mars was performed during the Viking missions. The elemental composition was determined by X-ray fluorescence, which analysed the composition for elements heavier than Mg. It was shown that the most essential major elements that make up biological matter, such as C, H, O, N, P, K, Ca, Mg and S, are present on the surface of Mars. However, as indicated above, organic compounds were not detected, as has been discussed by K. Biemann and others.

Indirect fingerprints of life

During its more than 3.5 billion years of history, life on Earth has substantially modified the terrestrial landmasses, oceans and atmosphere. Examples are the fossil deposits of petroleum and coal, the sediments of shell limestone, the coral reefs and the deposits of banded iron formation, biomineralisation and bioweathering, which all bear witness to biological activity in the geological past.

M. R. Walter has shown how stromatolites and related build-ups are examples of microbial biomineralisation, which are widespread in the geological record.

The composition and dynamic cycles of the Earth's oceans and atmosphere are decisively influenced by the biosphere. Examples are the water, CO_2 and nitrogen cycles. The absorption of light and heat from the Sun is also modified by the surface vegetation. Concerning the water cycle, evapotranspiration, especially of the tropical rain forest, is an important effect contributing to the release of water. The atmospheric O_2 is largely a product of photosynthetic activity that began in the early history of life, with cyanobacteria as the main producers. The break-up of oxygen molecules by sunlight in the upper layers of our atmosphere has led to the build-up of the UV protective ozone layer in the stratosphere. For the CO_2 cycle, the marine phytoplankton constitutes a large CO_2 sink, which is essential for maintaining the stability of the atmosphere.

Finally, there exist several biogenic minerals, which have distinctive crystalline structures, shapes and isotopic ratios that make them distinguishable from their non-biologically produced counterparts of the same chemical composition. Those minerals that result from genetically-controlled mineralisation processes and that are formed within a preformed organic framework have non-interchangeable characteristics, such as the orientation of the crystallographic axes and the micro-architecture. Examples are the skeletons of the unicellular marine micro-organism Acantharia, which is composed of strontium sulphate, the shells of amorphous silicate of diatoms, and the biogenic magnetite formed by bacteria, as described by D. E. Schwartz and others.

CONCLUSIONS

Before any 'search for extant life' experiments on a planet such as Mars, more data are required on Martian geology, including ancient lakes, volcanism, hydrothermal vents, carbonates; the climate, including the hydrosphere, the duration of phases that allow liquid water; the past and present morphological and biochemical signatures of extraterrestrial life; the utility of terrestrial analogues and the radiation environment, as well as organic molecules in sediments. The search for possible biological oases will focus on areas where liquid water still exists under present conditions. By analogy with terrestrial ecosystems, potential protected niches have been postulated for Mars, such as sulphur-rich subsurface areas for chemoautotrophic communities, rocks for endolithic communities, permafrost regions, hydrothermal vents, soil and evaporite crystals.

Much information can be obtained from remote-sensing global measurements, such as the seasonal atmospheric and surface water distribution, the mineralogical inventory and distribution, geomorphologic features obtained with high spatial resolution, thermal mapping of potential volcanic regions to determine possible geothermally active sites, and trace gases such as H_2, H_2S, CH_4, SO_x and NO_x.

In the case of Mars, it is also very important to explore and understand the strong oxidative processes on the surface. If the *in-situ* investigations confirm that the sites constitute potential biological oases, detailed searches for direct and indirect biomarkers should follow.

EVIDENCE OF EXTINCT (FOSSIL) EXTRATERRESTRIAL LIFE

Paleontological evidence

Given what we know about Earth's geology, the sedimentary record of a planet should serve as a store for both fossils of former organisms and other possible traces of their life activities. This should apply to Mars during the early stages of its history, when the planet had seen, at least locally, abundant water and environmental conditions on the surface that did not differ much from those on the early Earth, as described by Mike Carr and Heinrich Wänke. V. I. Moroz and L. M. Mukhin have shown how, if all terrestrial planets – and notably Mars and Earth – had occupied comparable starting positions in terms of solar distance, condensation history and primary endowment with matter from the parent solar nebula, then the surface conditions on both planets should have been very similar in their juvenile states.

Chris McKay and Carol Stoker have shown that, specifically, with evidence at hand for a denser atmosphere and extensive water activity on the Martian surface during the planet's early history, a convincing case can be made that the primitive Martian environment was no less conducive to the start-of-life processes and prolific microbial ecosystems than the surface of the ancient Earth. Indeed, author and physicist Paul Davies has even argued that the early Mars may have been even more conducive than the early Earth to the start of life (see below). J. W. Schopf and Manfred Schidlowski have discussed how, even if the evolutionary pathways of both planets diverged during their later histories, so that life became extinct on Mars as the gradual deterioration of surface conditions made the planet inhospitable, Mars could still have started off with a veneer of microbial (procaryotic) life comparable to that existing on the Archaean Earth. It could be argued, therefore, that with the apparent failure of the Viking life-detection experiment for the present Martian regolith, the prime objective of a search for life on Mars should be to seek evidence of extinct (fossil) life. The oldest Martian sediments would be appropriate targets for such efforts.

In any search for extinct life on Mars, the basic problem will not be recognising the fossil evidence, but rather with identifying those rocks with suitable sampling material from the large variety of potential host rocks exposed on the planetary surface. Among the two-thirds of the Martian surface covered by rocks older than 3.8 billion years, there are several occurrences of well-bedded sediments, notably in the Tharsis region and the associated Valles Marineris canyon system. These are interpreted as lake deposits and believed to include thick sequences of carbonates, as has been discussed by Chris McKay and S. S. Nedell. Sedimentary rocks have been discovered by the opportunity rover in meridiani planum as described by S. Squyres.

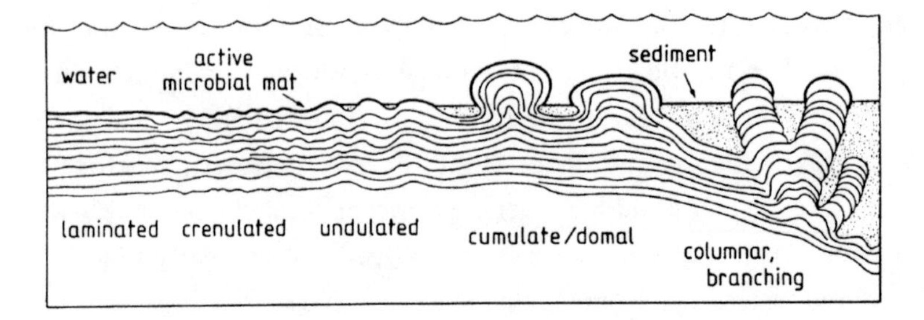

FIGURE 5.3 Scheme of principal morphologies of laminated microbial ecosystems that thrive at the sediment-water interface. They become fossilised in the form of 'stromatolites' (see Plate 8). The mat-forming is mostly made up of cyanobacteria.

If present at all in these early Martian formations, fossil life should lend itself to as ready a detection as its Earthly counterparts, provided the host rock is accessible to either robotic sensing or to direct investigation following a sample return mission.

To see how to investigate this, we note that the oldest sedimentary rocks on Earth are 3.5 to 3.85 billion years old and this may guide us as to how to look for fossil evidence from Martian rocks of the same age. On Earth, the earliest procaryotic (bacteria and archaea) microbial ecosystems have basically left two categories of evidence, namely (i) stromatolite-type biosedimentary structures called microbialites and (ii) cellular relics of individual micro-organisms or microfossils.

Microbialites

Microbialites or 'stromatolites', described by R. V. Burne and L. S. Moore, are laminated structures that preserve the matting behaviour of the original bacterial and algal growths. Microbial build-ups of this type are stacks of finely laminated fossilised microbial communities that originally thrived as organic films at the sediment-water interface, with younger mat generations successively superimposed on the older ones (Figures 5.3 and Plate 8).

On Earth, microbialites have a record extending back to the Early Archaean period of some 3.5 billion years ago. Studies of these stromatolites allow a fairly elaborate reconstruction of the Earth's earliest microbial ecosystems, and indicate that Archaean stromatolite builders were not markedly different from their geologically younger counterparts including contemporary species. M. R. Walter has described how the principal microbial mat builders were filamentous and unicellular procaryotes. The unbroken stromatolite record from the Archaean to present times shows the astounding degree of conservatism and uniformity in the physiological aspect and communal organisation of these procaryotes over 3.5 billion years of geological history.

Cellular microfossils

There is a second microscopic category of paleontological evidence stored in the sedimentary record. This microscopic or cellular evidence is currently supposed to go back to at least 3.5 billion years, as in the case of microbialites. However, the biological origin of microfossil-like structures reported from the ~3.8-billion-year-old Isua rocks from West Greenland found by H. Pflug has been discredited.

While a wealth of authentic microbial communities has been reported from Early and Middle Precambrian (Proterozoic) formations, the clear identification of cellular microfossils becomes notoriously difficult with the increasing age of the host rock. In Early Precambrian (Archaean) sediments, both the progressive diagenetic alteration and the metamorphic reconstitution of the enclosing mineral matrix tend to blur the primary morphologies of delicate organic microstructures. This results in a large-scale loss of contours and other critical morphological detail.

In spite of the evident impoverishment of the Archaean record, there are, however, single reports of well-preserved microfossils. Most prominent among these assemblages are chert-embedded microfloras from the Warrawoona Group of Western Australia, and from

6 After the discovery/life as a cosmic phenomenon

What if we do find life on Mars? The significance of the discovery will hinge crucially on whether Martian life is the same as terrestrial life. This is important, given the possibility that Mars and Earth have cross-contaminated each other. There is substantial traffic of material not only from Mars to Earth but also (since Earth occasionally suffers large asteroid and comet impacts, too) in the opposite direction – although fewer rocks go the other way, because of Earth's deeper-gravity well. Microbes may have been transported in either or both directions by this process. This intermingling of the two biospheres would considerably complicate the picture. It is entirely possible that life started on one planet and spread to the other before a second genesis could happen. It is a moot point whether introduced life would rapidly commandeer all available niches and food resources, thus stifling a second genesis, or whether two different biological systems could coexist on the same planet.

Paul Davies (2002)

Well-known theoretical physicist and popular author Paul Davies believes that life may well have originated on Mars. Since we are now confident that life (at least primitive forms) can be transferred from one planet to another, it would seem that terrestrial life could have had a Martian origin. Davies's argument is based on the fact that Mars is smaller than Earth and therefore cooled quicker and would have been ready for life as early as 4.4 billion years ago. This is contrasted with the Earth only being habitable about 3.9 billion years ago. There was a great bombardment of both planets by giant asteroids, and possibly comets, for about 700 million years after the formation of the Solar System some 4.5 billion years ago, and this created enormous traumas with extreme temperatures at the surface. With temperatures deep down in the planets high and only cooling slowly, the faster cooling on Mars could have allowed an earlier window of opportunity for life's emergence and survival on Mars rather than on Earth. If this is what really happened, then it would be logical to infer that some vestiges of that life are still present on the Martian surface or just below it.

AFTER THE DISCOVERY

What about the discovery itself and its aftermath? The discovery could fit three possible scenarios. The first is that the life form is very similar to that of a terrestrial one, strengthening the possibility of an interplanetary transfer of life and weakening subsequently the evidence of a true second genesis of life. The second is that the life form is different from anything we know on Earth, but still based on the DNA/RNA/protein scheme. The third could reveal a life form completely alien to our DNA/RNA/protein-based life we know on Earth, but still probably exhibit those characteristics which we have identified as signifying life, i.e. an organic system which is capable of self-reproduction and evolution.

The discovery of life on another planet will not be the endpoint of the search. Rather it will be the starting point of a whole new area of scientific research.

Concerning the discovery itself, and once the clear and unambiguous identification of life or traces of life is made, a number of issues will crowd to the front. These are issues related to:

- The geographical location
- The geological context
- The biological context
- The geochemical context
- The water context
- The ecological context
- The analysis context
- The human/robotic context
- The contamination context

The geographical location

This concerns the overall geographical location of the discovery, such as: on Mars is it in the northern or southern hemisphere? Is it in or outside a crater? Is it in a mountainous or plain region? Is it in or near cliffs? What altitude is it at and what are the corresponding atmospheric pressure and climatological conditions? Is it in equatorial, mid-latitude or polar regions? If the discovery has been made by robotic exploration during human mission/s, how easily can it be accessed by the human crew?

The answer to most of these questions should give strong hints as to where new searches should be directed.

The geological context

This is an extremely important issue. Questions such as what type of regolith or rock (igneous, metamorphic or sedimentary) is involved? Is the object retrieved from the surface or deep within the rock or regolith? Is the surrounding material recently exposed or long-term buried? Again, the answers to these questions will give clues as to where to look further, as well as some hints – if it is fossil material – about the age of the fossils and the time of deposition.

The biological context

This covers those aspects of the context specific and particular to biology. For example, if the discovery is of extant or fossilised micro-organisms, are they present in large numbers in colonies or are they sparsely placed in the sample? Is it possible to identify cellular elements, such as nuclei, cell walls, mitochondria etc.? Do they co-exist in the sample with other biomarkers or organic material? If the micro-organisms appear to be alive or recently alive, is it possible to determine anything about their metabolism? Are they extremophiles existing in extreme conditions of, for example, salinity, acidity/alkalinity, low or high temperature? If the micro-organisms are present in colonies, do these colonies appear to have modified the surrounding geological features? Are there any similarities with terrestrial organisms? The answers to these questions would go a very long way to answering the question about the uniqueness or not of terrestrial or for that matter Martian organisms.

The geochemical context

This relates to the chemistry and mineralogy of the sample in which the life object is found; in particular, the organisms that are present in the surroundings, as well as in the minerals. Of special interest is the possible presence of existing or reducing inorganics and the isotopic abundance ratios of the inorganic materials and minerals. From a biochemistry point of view the presence of typical terrestrial metabolites, such as ATP and substances involved in, for example, photosynthesis, would be significant. Also the presence of proteins known to be cell-wall receptors or transmitters would tell a lot about the interactivity of these cells with other cells.

The water context

This is extremely important and relates to the extent to which liquid water, either past or present, has been associated with the life markers that have been discovered. If it is extant life, say micro-organisms, we will need to know how close by is there liquid water and, if not close, how did the micro-organisms survive? At the back of it is the question of how important is the presence of liquid for the genesis and survival of extant micro-organisms? Has the case been overstated or not?

In the case of extinct life, the finding of fossils of micro-organisms will need to be correlated with the presence of past signs of liquid water in the vicinity. In particular, channels, outflows and traces of the movement of liquid water will be fundamental. Also sedimentary rocks formed by the sedimentation of minerals in a water ocean may be key areas where fossils are found. In addition, there is the interest of ancient shorelines, either oceanic or shores of lakes in, for instance, ancient craters, which might be key areas to find signs of ancient life. We know that life on Earth and its origins has many connections with the interface between land and ocean. It is known that the tides on Earth caused by the gravitational action of our Moon have played a significant role in the development of life on our planet. Liquid water and its role is crucial here.

The ecological context

There is not much to say here; either there will be biology present or not. However, we must be mindful that life is a complex phenomenon, often involving complex interaction between different species or organisms. It seems unlikely that we would find just one species present in that environment. The discovery of one species therefore would call for an immediate search for others in the vicinity, and study of the possible interactions between them.

The analysis context

This relates to the analysis procedure which has been used to determine the signatures of life. This will be concerned with how direct or indirect is the life inference. For instance, the finding of a DNA/RNA fragment by a smart microchip-based instrument would have a far stronger impact than a gas chromatograph/mass spectrometry detection of the basic chemical constituents of DNA/RNA.

The human/robotic context

This is fairly self-evident. Discoveries made by robots will need to be carefully assessed by human controllers/operators. Discoveries reported by humans in the field will also need to be verified by the mission control. Whereas robots may not be able to fully determine geological context, humans suitably trained as geologists would be expected as a matter of course to log all this.

The contamination context

This is one of the most important aspects. Any discovery will have to be ascertained in a way which ensures that there can be no question that the life or life signatures found are the result of forward contamination, brought either by the humans or by the robotic missions which precede the human ones. This point is absolutely crucial. Any hint or suggestion that the discovery could be of signatures brought from Earth would destroy the credibility of the claim. Strenuous efforts will need to be undertaken to obviate this risk.

The 'objects' discovered could be the following:

- living micro-organisms
- fossils of micro-organisms

- viruses
- prions
- fragments of DNA or RNA
- antibodies
- other genetic type sequences
- kerogens or other biomarkers
- cellular components, e.g. cell membranes, mitochondria
- genes or whole chromosomes

With the exception of the kerogens and biomarkers there should be some chance to investigate the genetic machinery of the organism or remnant. Even if Davies' theory is right, and life started on Mars and then moved to Earth, the sequence of questions to be answered will not be different and probably will be as follows:

- Determine if the information-carrying mechanism is the DNA/RNA one
- If not, determine what it *is* based on. A scientific revolution would follow
- If yes, determine the DNA genome of the Mars micro-organism
- Compare it with genomes of Earth micro-organisms
- Determine what is new/unusual
- Determine the ethical steps needed to continue the research

If extant micro-organisms are found, a number of question paths will present themselves as follows:

LIFE AS A COSMIC PHENOMENON

It is impossible to understate the impact of a clear and verified discovery of autonomous life forms on other planets of our Solar System. This would have impact not only on the daily news-chatter and preoccupations of Earth-bound humanity but also in the halls of serious science and philosophical communities, institutions and debate. Much intellectual capital has already been invested in both sides of the argument of whether or not life is unique to our planet. In general,

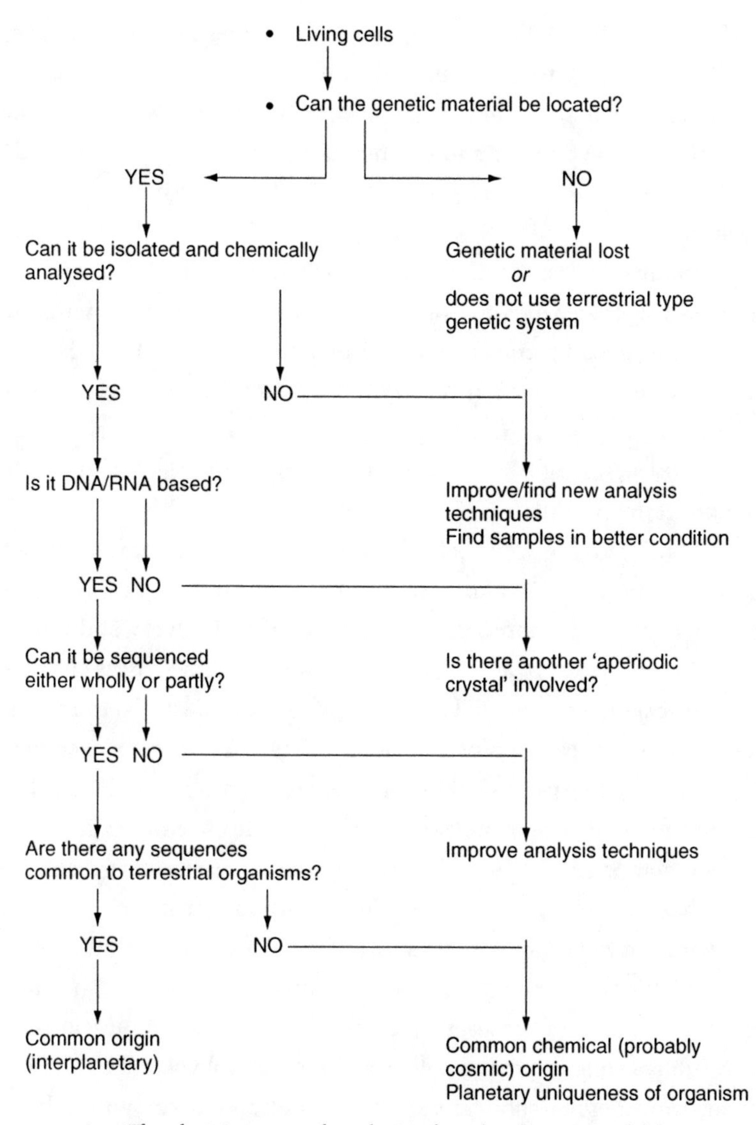

FIGURE 6.1 The decision tree of analysis after the discovery of life in a non-terrestrial context.

with some noted exceptions, the literary/humanities axis dismisses even the discussion as nothing more than the preoccupation of science fiction buffs. The scientific and technology communities are probably more evenly split, but still a good 50 % (probably most of the astronomy, high energy physics and other physical sciences communities) tend to be hostile to the idea. The most open discussions are between planetary scientists, biologists interested in self-organising systems and the organic and biochemists in the Urey-Miller mould who would like to create life or at least its basic constituents in the laboratory.

So the discovery would settle all that and cause a real sensation. Witness the newspaper and media fuss when the Martian meteorite story broke, accompanied by momentous statements by US President Bill Clinton and other politicians. The Allan Hills meteorite story was, of course, with hindsight, a premature announcement of an apparently significant but, in the end, inconclusive piece of evidence for extraterrestrial life.

The discovery would also have profound consequences for our view of biology that is sometimes said to have superseded physics as the scientific powerhouse of modern scientific discovery and innovation. The discovery of new organisms, new genetic sequences and new metabolic pathways would lead not only to the almost instantaneous birth of new armies of biologists, biochemists and genetics researchers to study the scientific and philosophical consequences of all this, but also to the birth of new biotechnology industries keen to explore the commercial possibilities for new pharmaceutical products resulting from these discoveries. It is often forgotten that most new drugs on Earth are derived from plants resulting from Earth's rich biodiversity. It is difficult to imagine that new life forms found in space, or more likely on planetary surfaces, would not have potential implications for such innovations in pharmacology and biotechnology and the downstream commercial applications that would undoubtedly flow from them.

The extent to which our view of the Universe and our place in it has changed over the last half-century is enormous. In 1957, before Sputnik bleeped its way across our skies, the concept of 'outer space' was an extremely abstract notion for most, even educated, people. The idea of there being organic molecules, the building blocks of life, in

FIGURE 6.2 Yuri Gagarin prepares to orbit the Earth in Vostok 1 in April 1961, the first human in space. (Courtesy NASA.)

this interplanetary 'outer space' was fiercely resisted by many in the scientific community well into the 1970s, despite pioneering work by Fred Hoyle and others. The courageous voyage of Yuri Gagarin as the first man in space, and the profound political and visionary imagination of President John F. Kennedy when he announced the Apollo Programme transformed the way we see our planet and its Moon.

The realisation of life as a cosmic rather than a terrestrial phenomenon will probably have little revolutionary impact within that part of the scientific community that already accepts it. But its impact on the thinking of literary, philosophical and political opinion formers may well be enormous.

Part III
Life in the search for life beyond the Earth

In reporting my predicament, certain members of the press have described the South Pole as 'hell on earth'. They refer to my time on the Ice as 'an ordeal'. They would be surprised to know that here, in this lonely outpost surrounded by the staggering emptiness of the Polar Plateau, in a world stripped of useless noise and comforts, I found the most perfect home I have ever known.

Dr. Jerrie Nielsen (2001)

...there are limits in their [probes] powers of observation. They can't readily adjust to fleeting conditions; for example, how do you tell a robot to look for a plankton bloom or watch for a unique flash of lightning? This is why [scientists] crave astronaut observations... even though they have access to robot-taken photos.

R. Mike Mullane (1997)

Decades after the Viking biology experiments, for example, there are still arguments as to whether or not they detected bacterial life. A scientist on the surface could have found the truth in a few hours.

Clark S. Lindrey (1999)

7 The prospects for long-duration human spaceflight and human survival on planetary surfaces

GENERAL HEALTH ASPECTS AND RELIABILITY REQUIREMENTS

When the *Beagle* with its crew of more than 60 left England in 1831 to sail around the world, the crew would have been completely engaged in preparations for the voyage and for the discoveries they expected to make during this voyage. Probably nobody calculated that after nearly 5 years, when the vessel returned to England, most of the crew would have experienced several more-or-less severe illnesses. Whenever a group of people goes on a long-duration expedition, some of them will become ill or even die. This happens regardless of whether the group stays at home or not. The individual risks of illness, injury and death can be deduced from general human health statistics. Time is the main factor here: the longer the duration, the more likely it is that individuals will fall ill or die. Added to this are the risks inherent in the expedition itself.

For human space missions, certain limits of acceptability have been set by space agencies with regard to the risks of illness, injury or death. They are called reliability objectives and are defined as follows: (i) the individual risk of death from illness during a mission shall not be greater than 0.2 % per year; (ii) the individual risk of death from injury (excluding spacecraft failure) during a mission shall not be greater than 0.04 % per year; and (iii) the individual risk of death from all causes, including spacecraft failure, shall be maintained during the mission at less than 3 % per year. The first two values are taken from the statistics of 'natural death' or 'accidental death' for the general population in Western countries, the latter value being generally accepted for human activities in professions with high risks, such as fighter pilots, helicopter pilots and – as in this case – astronauts. Although for each human space mission these reliability requirements have been met, the nature of the accidents so far experienced in manned spaceflight has shown that the most critical periods for spacecraft failures are at launch and re-entry. Examples, of course, are the Challenger explosion in 1986, killing a crew of seven about one minute after launch, and the Columbia break-up upon entering the Earth's atmosphere on 1 February 2003, killing another crew of seven. It is important to understand that the reliability objectives are based on statistics. Statistics suggest the probability that an event may happen, but nothing about whether and when it may actually happen.

It can be anticipated that the above reliability objectives will also be applied by agencies for lunar base scenarios and human Mars missions. For designing and sizing the crew health control system for a 180-day lunar mission, a probability of death by illness and injury (excluding spacecraft failure) for the whole crew was calculated in the HUMEX study to be less than 0.02 % aboard the Earth-Moon transfer vehicle, and less than 0.47 % onboard the lunar lander and in the lunar habitat. For the long-term Mars missions, these values would be less than 0.17 % for the Earth-Mars transfers and onboard the Mars orbiter and 2 % onboard the landing module, respectively. It is important to note that all these reliability values calculated for the lunar and Mars scenarios are compatible with the mission safety objectives outlined above.

These reliability objectives also assume that there is a significant probability of diseases and injuries occurring during any long-term expedition, such as the Mars exploration missions. Examples are infectious diseases, cardiovascular diseases, digestive and dental disorders, as well as other health problems to the eye, ear, nose and throat; in fact, just the same catalogue of disorders which will occur in any group of people over a time period of more than three years, even if especially healthy individuals are selected. On top of this catalogue, account must be taken of both minor and serious injuries, such as bone fractures, crushing, burns and open wounds. This means major challenges for the life sciences and health control systems on board, especially in view of the fact that an emergency return is impossible, and that communications with Earth may be delayed by up to 20 minutes out and back. Therefore, a real-time telemedicine system will not be a suitable solution for Mars missions. Human exploratory missions such as those to Mars will require sophisticated, compact and highly autonomous systems for diagnostic and therapeutic purposes needed to treat all the most common diseases.

LIVING AND WORKING IN A WORLD WITH REDUCED OR ZERO GRAVITY

Life on Earth, since its emergence about four billion years ago, has evolved under the constant force of gravity. Gravity, on the one hand, is a factor of physical constraint forcing life to counter it, for example, by developing different supporting structures, such as muscles and bones in animals or fibres and wood in plants, in order to overcome gravity-enforced size limits or to maintain the shape of living things. On the other hand, since gravity is continuously present and has a fixed direction, it has been used by various forms of life as a valuable external signal for orientation and postural control purposes. Throughout its history, life on Earth has never been confronted with a reduced or vanishing gravity vector and hence both orientation behaviour and ontogenetic development are thoroughly adapted to the Earth's gravity.

This situation changed with the beginning of spaceflight. The first animals in space, like the famous dog Laika, demonstrated that life can continue to function under conditions without gravity. So far, hundreds of humans have spent days, weeks, months or in some cases more than a year in low Earth orbit. The adaptation processes of life to weightlessness have been thoroughly studied, and several countermeasures have been developed to cope with the deleterious effects resulting from extended exposure to weightlessness.

During a journey to Mars, the crew will experience weightlessness over an extended period of time, longer than ever studied in Earth orbit. This will be especially severe for the two crew members who remain in Martian orbit and experience a total period of more than

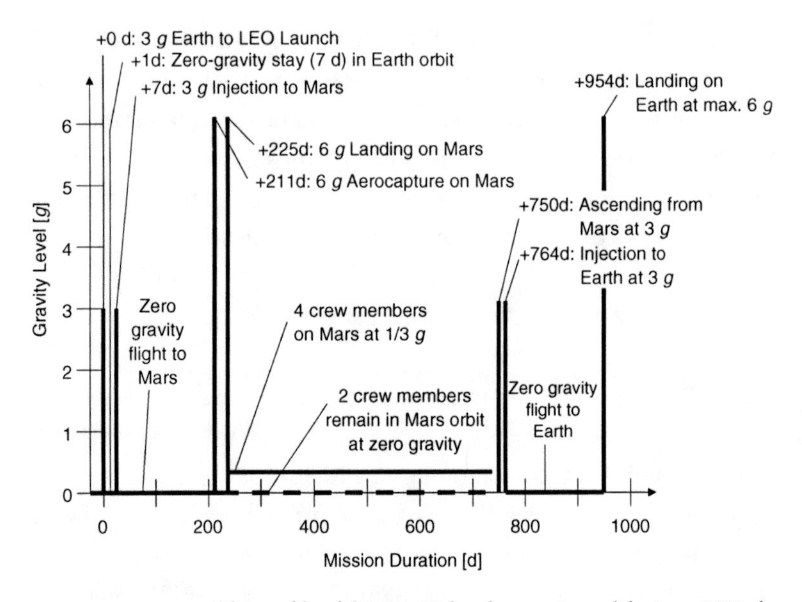

FIGURE 7.1 The profile of the gravity levels experienced during a 1000-day Mars mission (from HUMEX study).

3 years of weightlessness. In addition, the crew will experience various transitions through different *g*-values, starting from 1 *g* on Earth through hypergravity during launch to weightlessness (also called microgravity), followed by a long-term exposure to microgravity during the interplanetary transfer, and then, upon arrival at Mars, an abrupt transition from microgravity to hypergravity with heavy *g*-loads of up to 6 *g* maximum during aerocapture and landing manoeuvres. Then 4 members of the crew will be exposed to the reduced Martian gravity of 0.38 *g* for more than a year (Figure 7.1). And this is just one half of the round-trip time between the Earth and Mars. At present, not very much is known about the responses of the human body to these switches between different levels of the gravitational force.

During the first days in space most of the crew experience the so-called space adaptation syndrome. This results from the neurosensory and cognitive responses to the transfer from 1 *g* to weightlessness, and causes symptoms similar to sea sickness or motion sickness, i.e. nausea and vomiting. Although these adverse consequences can largely be controlled pharmacologically, the performance of the crew will be impaired to a certain extent and activities will be quite limited during this adaptation period. However, unlike sea sickness, which may last for the whole sea journey, the space adaptation syndrome disappears after a few days, when the neurosensory system has adapted to the new gravity conditions.

Charles Darwin would have envied the astronauts for this short adaptation period. He stated after completing his voyage with the *Beagle*, that he would well have undertaken another voyage of discovery if it had not been for the chronic sea sickness which he suffered. He wrote:

> *I have been paying the Beagle a visit today. She sails in a week to Australia. It appeared marvellously odd to see the little vessel and to think that I should not be one of the party. If it was not for the sea sickness, I should have no objection to start again.*

Although this motion sickness in space disappears after 3 to 4 days, a similar adaptation syndrome will probably occur with every change of the gravity value, as has been observed with astronauts returning to the Earth's gravity from weightless conditions in low Earth orbit, such as on the ISS. Nothing is known about the neurosensory adaptation to the reduced gravity environment of 0.38 *g* on Mars. So far, the only data available on transitions from weightlessness to a reduced gravity are from the Apollo lunar missions. In the clinical reports of the 5 lunar missions (Apollo 12, 14, 15, 16 and 17), R. S. Johnston and others found that no phenomena similar to motion sickness were reported by astronauts after landing on the Moon's surface.

As mentioned above, the long-term exposure to weightless conditions during the transfer phase from Earth to Mars and back, as well as in Mars orbit – the latter at least for part of the crew – will have major implications for the astronauts' health. Substantial information on the effects of long-term weightlessness on human performance and health has been gained from extended studies onboard the MIR and the ISS. Deconditioning symptoms have been observed, such as loss of

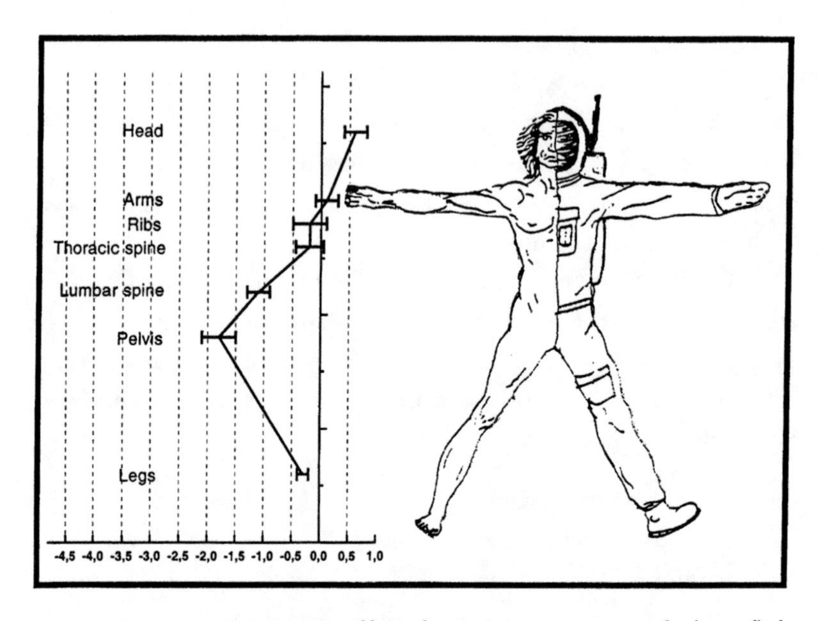

FIGURE 7.2 Loss in mineral bone density in per cent per month of spaceflight weightlessness (from HUMEX study).

muscle and bone mass, reduced cardiovascular and physical capacity, and changes of motor skills; although during any journey to Mars, crews will experience weightlessness over periods of time longer than those studied up to now.

Under weightless conditions, a continuous loss in mineral bone density has been observed; however, the extent varies between different parts of the skeleton (Figure 7.2). The most severe bone loss has been observed at the level of the pelvis with up to 2 % loss per month of spaceflight, which has been measured over a period of 6 months on board the space station MIR. The values at the level of the femoral neck indicate a loss rate of 1.2 % per month and at the level of the femoral trochanter of 1.6 % per month during spaceflight. Extrapolating from these results to a 1000-day Mars mission, the level of bone density loss could reach 38 % at the femoral neck and up to 50 % at the femoral trochanter. These levels far exceed the threshold for significant risk of fracture that is set at an of amount of bone loss of 15 %. For that reason, bone demineralisation during a 1000-day space mission still remains an unacceptable and uncontrolled risk. These changes in bone mass are comparable to the physiological deconditioning phenomena observed in elderly people. However, in space the bone mass loss is three to ten times greater than the loss associated with ageing on Earth.

This amount of bone loss can be life threatening, especially in emergency situations, or during EVAs, and while on a planetary surface. These losses occurred, although the cosmonauts and astronauts on board the MIR station from whom the data were obtained had performed physical exercise twice per day and/or worn a so-called 'penguin suit' in order to counteract unfavourable weightlessness effects. Therefore, more efficient countermeasures against the negative effects of weightlessness need to be developed.

Under weightless conditions, physiological deconditioning affects nearly all body functions. In addition to the loss of bone mass, a loss of muscle mass has been observed, mainly in the legs and trunk, with a rate of 0.6 % per day in the soleus. This may lead to a decrease in the muscle strength and an increase in muscle fatigue, eventually resulting in changes in motor control. The data show that the astronauts who had exercised during spaceflight had less muscle impairment after flight than those who had not.

Another phenomenon is the so-called 'puffy face' caused by a fluid shift in the human body towards the upper torso and head, because there is no gravity to pull it down to pool in the lower extremities. As an adaptation to the weightless environment, the body responds with a reduction of the plasma volume. When returning to 1 g conditions, the reduced body fluid volume contributes to orthostatic intolerance, or difficulty in standing and balancing. However, this post-flight orthostatic intolerance is reversed within a few days after landing, even after a six-month spaceflight. This orthostatic intolerance can be mainly overcome if the astronauts apply appropriate countermeasures prior to landing. For long-term missions, it is advised to wear the 'lower body negative pressure device' for certain periods of time during the last 10 days of the spaceflight in order to counteract the fluid shift caused by weightlessness. In addition, wearing of the anti-*g* suit during return and for the first few days in gravity is recommended. The rehabilitation process on ground, lasting for about 4 days, consists of appropriate salt and fluid intakes, and in the progressive releasing of the anti-*g* suit under daily stand tests. Similar countermeasures could be effective when crews arrive at the 0.38 *g* environment of Mars, although this aspect needs confirmation. So far, the only data about changing from weightlessness to reduced gravity levels are available from the lunar Apollo missions. For these missions, no orthostatic intolerance was reported during the Moon surface exploration.

Whereas the loss in bone mass continues with the duration of spaceflight, most other physiological adaptation phenomena reach a maximum after a certain period of time, usually within one to two months, and then return more or less to a steady state, in most cases close to the level in 1 *g* conditions (Figure 7.3). Therefore, loss in mineral bone mass – besides the radiation effects that will be discussed later – turns out to be one of the major risk factors connected with long-term exposure to weightlessness during missions to Mars. Hence, it is of utmost importance to develop suitable countermeasures against the effects of weightlessness or reduced gravity. Intensive exercise, such as the intense use of a treadmill, have been used to overcome the negative effects of weightlessness on the physiological systems of the body. However, so far, even with a very rigorous exercise programme carried out on the MIR station, neither the bone density nor the integrated muscle functions were at Earth levels upon return. Therefore, additional countermeasures to cope with the negative effects of weightlessness need to be developed and tested. The ISS is a good test bed for such studies.

In several instances – and not only in the science fiction literature – artificial gravity has been considered as an alternative means for preventing weightlessness-induced disorders, such as bone loss, cardiovascular orthostatic intolerance, muscle atrophy and exercise capacity impairment. This could be achieved either by two tethered sister spacecrafts rotating around a common axis or by the use of a rotating

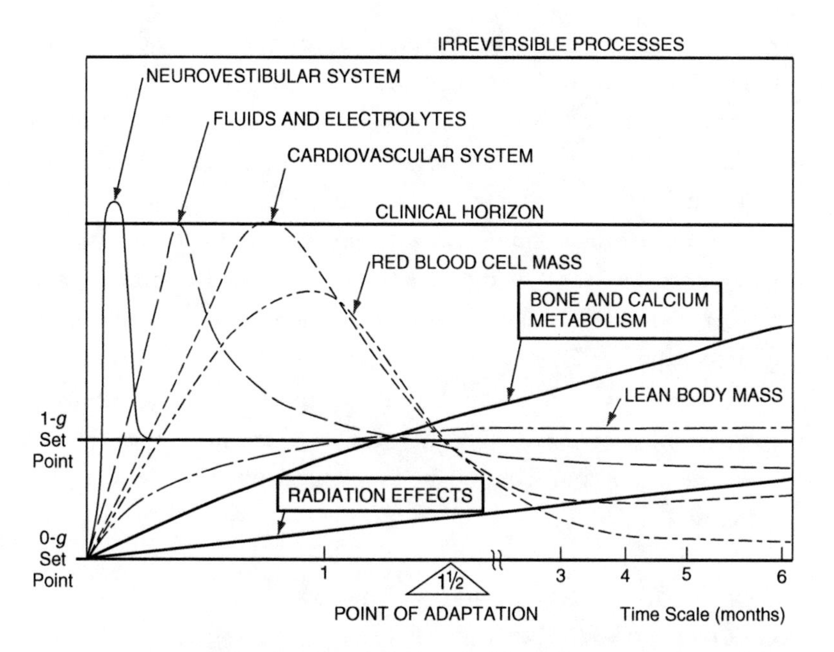

FIGURE 7.3 Development of physiological deconditioning symptoms during spaceflight (from HUMEX study).

spacecraft or a centrifuge. However, in order to obtain at least an artificial gravity of only 0.35 *g* within a comfortable zone, a minimum radius of 17 metres and a maximum angular velocity of 4 rpm would be required. A shorter radius would inevitably cause a gravity gradient within the human body, accompanied with Coriolis cross-coupled angular acceleration. Because of this large size required for a comfortable zone in artificial gravity and the low level of gravity reached in this zone, a permanent exposure to artificial gravity seems to be not practical at present for exploratory missions. It is also not clear whether or not 0.35 *g* would be sufficient to avoid the deconditioning symptoms of weightlessness discussed above.

Alternatively, the use of a transient artificial gravity at specific intervals during a mission has been considered as possible means to prevent, at least in part, negative spaceflight effects. To do this, a self-powered short-arm human centrifuge has been developed. However, so far, experiments with this short-arm centrifuge alone did not show any benefits on the maintenance of exercise capacity. In addition, the experiments showed clearly that discomfort occurred due to bouts of motion sickness as a consequence of a changing *g*-force. More experiments on board the ISS are required to validate the benefits of a short-arm centrifuge, and to test whether the astronauts will be able to adapt to the stress of different gravitational environments and whether the side effects of these transitions will diminish over repeated exposures. Although one might tolerate bouts of motion sickness associated with Coriolis cross-coupled angular acceleration early in a mission, it would clearly not be acceptable if these symptoms were to reoccur each time the astronaut entered different gravity environments.

LIVING IN A RADIATION FIELD OF INTENSITIES NOT EXPERIENCED ON EARTH

Cosmic radiation, of solar as well as of galactic origin, is a major source of energy arriving at the biosphere. The effects of the heavy ions of cosmic radiation on life on Earth is less clear than the obvious effects of the Sun's light on our biosphere. This is because the surface of the Earth is largely shielded from the cosmic radiation, due to the deflecting effect of the geomagnetic field and the huge shield of $1000\,g/m^2$ provided by the atmosphere. As a consequence, the cosmic component of the radiation dose at the surface of the Earth is reduced by about a factor of 100, compared to that in low Earth orbit.

Since the advent of human spaceflight and the establishment of long-duration space stations in Earth orbit, the upper boundary of our biosphere has been extended into space. Humans and other biological systems in space are exposed to a radiation environment of a composition and intensity not encountered on Earth. In interplanetary space, the radiation field is composed mainly of two groups: (i) the solar cosmic radiation (SCR), and (ii) galactic cosmic radiation (GCR). In the vicinity of the Earth, a third radiation component is present: the radiation trapped by the Earth's magnetic field, the radiation belts.

Solar cosmic radiation consists of the low energy solar wind particles that flow constantly from the Sun, and from the so-called solar particles events (SPEs) that originate from magnetically disturbed regions of the Sun that sporadically emit bursts of charged particles with high energies. These events are composed primarily of protons, with a minor component (5–10 %) being helium nuclei (alpha particles) and an even smaller part (1 %) heavy ions and electrons. SPEs develop rapidly and generally last for no more than some hours; however, some proton events observed near Earth may continue over several days. The emitted particles can reach energies up to several giga electron-volts (GeV). Such strong events are very rare, typically fewer than three events occur during the 11-year solar cycle. For low Earth orbit (LEO), the Earth's magnetic field provides a shielding against SPE particles. This shielding is decreased at higher latitudes however, so that in high inclination orbits or during interplanetary missions, SPEs can pose a hazard to humans in space, especially during extravehicular activities.

Galactic cosmic radiation particles originate outside the Solar System. They have their origin in cataclysmic astronomical events, such as supernova explosions. The particles consist of 98 % baryons and 2 % electrons. The baryonic component is composed of 85 % protons (hydrogen nuclei), with the remainder being alpha particles (helium nuclei) (14 %) and heavier nuclei (about 1 %). The particles come from all parts of the sky and can have energies up to 10^{20} electronvolts.

The radiation belts are a result of the interaction of the GCR and the SCR with the Earth's magnetic field and atmosphere. Two belts of radiation are formed, comprised of electrons and protons, and some heavier particles trapped in closed orbits by the Earth's magnetic field. The main production process for the inner belt particles is the decay of neutrons produced in cosmic particle interactions with the atmosphere. The outer belt consists mainly of trapped solar particles. In each zone, the charged particles spiral around the geomagnetic field lines and are reflected back between the magnetic poles, acting as mirrors. Electrons reach energies of up to 7 MeV and protons up to 600 MeV. The energy of trapped heavy ions is less than 50 MeV. The trapped radiation is modulated by the solar cycle: proton intensity decreases with high solar activity, while electron intensity increases, and vice versa.

The different types of radiation in space cause different amounts and kinds of biological damage per unit of absorbed dose. Therefore, the different types of radiation received in space have to be known and accounted for. Measurements have shown that the total dose rate is about twice as high for the MIR 92 mission as for the IML-1 Spacelab/Shuttle mission, and the contributions of the different components of radiation to the total dose are quite different for the two missions. The reason for this difference is the higher altitude of the MIR orbit compared to that of the IML-1 mission and a different period in the solar cycle. However, common to all missions in low Earth orbit – and thereby different from interplanetary missions – is the protection of the spacecraft by the Earth's magnetic field against the randomly occurring SPEs.

The exposure profile to ionising radiation of galactic and solar origin experienced by the crew during a mission to Mars is shown in Figure 7.4. Because the mission scenario does not encounter trapped radiation, apart from the exposure incurred during the short crossings of the terrestrial radiation belts, this contribution may be neglected compared to the radiation levels experienced during the rest of the mission. A component which only affects radiation exposures indirectly is the solar wind, which modulates the primary galactic heavy ions according to the rhythm of solar activity, with an average cycle time of around 11 years. In order to assess the radiation exposure, the following parameters are also relevant: mission duration, solar distance, solar cycle time, mass shielding and mission phase.

Upon arrival at the surface of Mars, a certain shielding is provided by Mars itself, roughly reducing the dose of GCR by a factor of 2, and in addition by the thin atmosphere with a shielding of 5–16 g/cm^2, depending on altitude. An additional minor radiation exposure comes

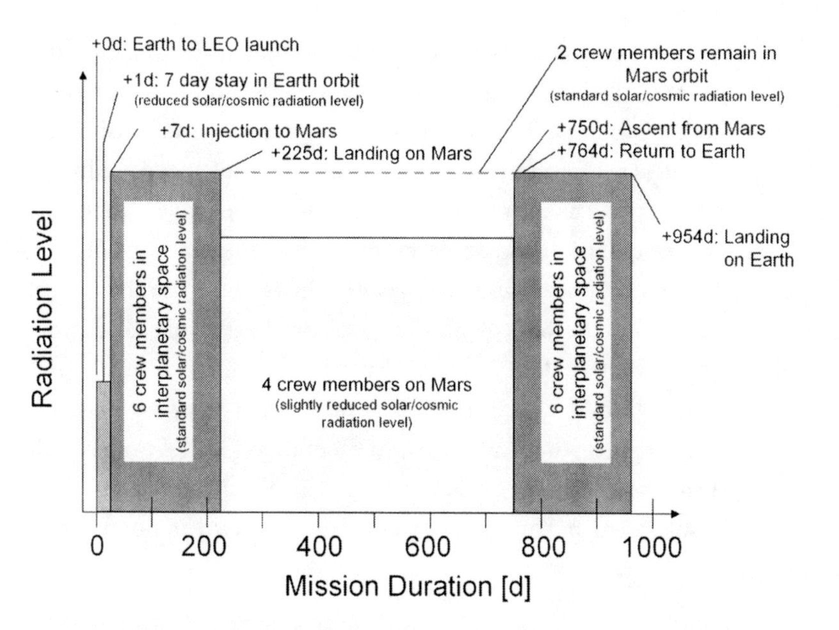

FIGURE 7.4 Profile of radiation levels experienced during a 1000-day mission to Mars (from HUMEX study).

from emissions from radioactive rocks on Mars itself. With increasing depth, the Martian soil provides a certain radiation protection. So far, no radiation measurements have been made at the surface of Mars. Our knowledge about the radiation conditions on the surface of Mars and their biological effects is based on a few radiation measurements made outside Earth orbit during the Apollo missions, reported by R. C. Johnston and others in 1975, and the dosimetric data obtained by the MARIE facility on board the Mars orbiting spacecraft Odyssey, launched on 4 April 2001.

Since the very beginning of human spaceflight, cosmic ionising radiation has been recognised as a key factor of the space environment, potentially limiting the duration of human stays in space by its deleterious effects on crew health, performance, and – finally – life expectancy. Therefore, for each type of mission, radiation protection guidelines have to be developed to enable humans to live and work safely in space.

Furthermore, if plants or animals are present during the mission, either as test systems or for bioregenerative life support purposes, knowledge of the effects of the space radiation on these inside the spacecraft is also of importance to safeguard the health and wellbeing of the crew, as well as the reliability of the bioregenerative life support systems. The radiation protection guidelines are mainly based on data obtained from dosimetry and modelling of the radiation field in space; from studies on the biological effects of the heavy ions of cosmic radiation encountered in space or produced at heavy ion accelerators on the ground; and from studies on the potential interactions of cosmic radiation with other parameters of spaceflight, above all weightlessness.

Exposure to radiation in space results in two main types of risk:

- Long-term exposure to expected levels of SCR and GCR resulting in an enhanced probability of cancer, and, possibly, changes in the cells of the brain, blood, reproductive organs and other tissue, occurring up to 20 or more years after exposure.
- Short-term consequences of relatively high levels of radiation, caused by the eruption of SPEs. This latter type of radiation risk is restricted to interplanetary missions and does not occur in low Earth orbit. It may cause severe cell depletion of sensitive tissues, such as bone marrow, intestinal epithelium, skin and others, and may lead to acute symptoms affecting the health and performance of the crew during the mission.

For missions in low Earth orbit, such as at the orbit of the ISS, the radiation protection guidelines recommended by the US National Committee of Radiation Protection take into consideration the gender and age of the individual. These age- and gender-dependent whole body career limits have been developed with the aim of keeping the radiation-induced late cancer mortality below 3 %. They are further enhanced by the ALARA principle, i.e. the requirement to keep the dose 'as low as reasonably achievable'. During missions through interplanetary space to other planets, such as Mars, as part of humankind's way outward into the cosmos, the safety provided so far in LEO has to be replaced by adequate means and measures available and effective within the spacecraft or habitat itself.

Nominally, exposures in most of the circumstances occurring in missions to Mars will comply with the radiation limits recommended for the ISS for females above 45 and males above 35 years of age – especially if a certain shielding is assumed. However, the constant exposure to elevated radiation doses during more than 3 years may increase the resulting cancer mortality beyond the design value of 3 %. For a mission during solar maximum conditions, any shielding thickness would keep the average dose rate low enough, because the interplanetary magnetic field protects objects in the Solar System better against GCRs than during solar minimum conditions. These values would strongly favour missions during periods close to maximum solar activity. On the other hand, this choice would significantly increase the probability of encountering large SPEs, which accumulate at the end of a solar maximum. The corresponding doses deposited by the worst case reference event in deep space, could possibly induce acute radiation injuries even behind a shielding of 5 g/cm^2 aluminium, and even if the crew has retreated to a radiation shelter providing a shielding of 10 g/cm^2 aluminium.

In the less protected surroundings on the surface of Mars, especially if the crew is caught by an SPE when exploring the surface of Mars during an EVA, the dose from one such event would surpass the total permissible mission dose. In this case, severe and even incapacitating acute radiation injuries could ensue, with a substantial probability for a fatal outcome unless adequate medical support could be supplied. To cope with such hazards, a reliable alert system is required to allow the astronauts to reach protective shelters in case of an SPE. In this case, the crew can benefit from the fact that the charged particles, above all protons, travel slower than light, so that there is a lag of many hours between seeing the SPE and the arrival of the protons.

In order to reduce the risks from radiation, three classes of strategies can be employed:

i. careful planning of mission duration, timing and operations;
ii. surrounding the crew habitats with sufficient absorbing matter; and
iii. increasing the initial resistance of exposed personnel to the deleterious consequences of exposure.

Since radiation risk is an increasing function of mission duration, the most effective countermeasure is to restrict exposure times to the absolute minimum. A full-blooded application of this principle may well warrant drastically reducing mission duration by developing and employing alternative propulsion systems. To the extent that other potentially limiting factors, such as medical/physiological problems from long-term exposure to microgravity or psychological problems from long-term confinement and seclusion, might even increase super-linearly with mission duration, developing more potent propulsion systems in order to reduce mission duration will be the countermeasure with the optimal cost-benefit ratio in the longer term.

Reducing the radiation risk by choosing – within the constraints imposed by celestial mechanics – the timing of a mission within the cycle of solar activity has to be balanced against other constraints. Exposure to GCR – with its large but still not well-known radiobiological effects – is at a minimum during phases of maximum solar activity, which clearly would favour such a choice. On the other hand, high solar activity implies increased probabilities for larger doses from SPEs. In particular, the giant SPEs so far observed always occurred during or shortly after the maximum of solar activity. Any quantitative risk/exposure trade-off between these alternatives requires the detailed specification of the complete mission with all its parameters.

Regarding shielding against space radiation, it is now known that due to a minimum build-up of secondary radiation products within the absorber, hydrogen instead of the conventional 'standard' shielding materials like aluminium, is the best choice, in particular for shielding against GCR. Hydrogen-rich solids like polyethylene or liquids like water are nearly as good. The integration of food or water into the shield is a possibility, but due to their eventual consumption they would not be available for the Earth-return phase.

On planetary surfaces, the mass shielding of habitats by surface materials is possible, provided the necessary construction facilities are at hand. However, this means additional mass and fuel for the transport of such construction equipment to Mars, for example. In any case, as a minimum, a shelter with sufficient shielding against SPE radiation will be needed for access before particles from an SPE reach the spacecraft or Mars.

Risk avoidance by the selection of crew members is also possible, and the choice of older crew members would reduce the chance of the loss of healthily lived years after the mission and could be exploited if these crew are reasonably physically fit. Although seldom considered explicitly, the use of crew of age beyond the usual human reproductive phase would also avoid the risk of later consequences in children conceived by the crew after the mission. In addition, gender differences in susceptibility to cancer would favour male crew members due to the added risk for females of breast cancer. This may have to be balanced with the requirement for mixed gender crews for psychological reasons.

There are also large individual variations of sensitivity to ionising radiation. Research in this area may eventually be useful for the identification and selection of crew members with increased resistance to early as well as late radiation effects.

However, if, in spite of all these carefully planned avoidance strategies, deleterious space radiation exposures occur, mitigation of the effects will have to rely on pharmacological measures, i.e. the administration of therapeutic and – preferentially – prophylactic radioprotective drugs. The performance of surgical therapeutic measures, such as bone marrow transplantation, hardly seems feasible in the present- and nearer-term planning of future Mars missions, even if highly sophisticated medical and surgical systems are planned to be available. However, most of the radioprotective chemicals so far known have severe side effects.

Research therefore will concentrate on naturally occurring biochemicals. Important among these substances are the antioxidative compounds, such as Vitamin A or E. Since many of these compounds occur in significant concentrations in natural food, they could serve as substances for space radiation protection if used in the diet during the mission. Whereas dietary control of tissue/cellular concentrations of enzymes and other proteins of the antioxidant defence system is limited, phytochemical antioxidants are affected more directly by the amount and composition of the diet and its supplementation with food derived from the respective plants.

LIVING IN A CONFINED HABITAT IN SPACE OR ON A PLANETARY SURFACE

Although the crew of missions to Mars will start the journey highly motivated, the daily monotony during the long interplanetary transfer phases will give rise to psychological problems. With increasing distance from the Earth, the blue home planet will become gradually smaller and smaller, first reaching the size of our Moon, and finally becoming nearly indistinguishable from the billions of other stars shining in the night sky. The Earth will become a pale blue dot, practically out of view of the Mars explorers. During this long voyage to Mars and the subsequent stay on the Martian surface, crew members will endure long periods of extreme confinement and isolation, which may cause severe psychological stresses.

The psychological stresses that astronauts and cosmonauts have experienced on board space stations, such as MIR, can be basically divided into four groups:

i. those particular to the space environment, such as weightlessness and alterations of the usual dark-light cycle;
ii. those related to the technical constraints of the space habitat and its life support system, such as confinement, limited facilities and

supplies for personal hygiene, elevated noise levels and elevated CO_2 concentrations in the air supply;

iii. those related to the specific operational and experimental workloads, such as work underload and overload and sustained stress; and

iv. those arising from the psychosocial situation in a space habitat, such as isolation from family and friends, lack of privacy and restricted and/or enforced interpersonal contacts.

In addition, cultural differences are an added factor in international space missions, influencing the emotional balance of the crew. Exposure to these different stresses has been shown to induce different behavioural stress responses in the individual astronaut or even the entire space crew. These can have three interdependent effects:

- impairment of cognitive performance and perceptual motor skills
- maladaptive individual behavioural reactions
- disturbances of interpersonal relationships within the space crew and between the space crew and the ground personnel

Several of the psychological problems observed on board space stations also arise whenever a small group of individuals live in an isolated environment. Hence, lessons on how to overcome psychological problems have been learned not only from the crew on board space stations but also from similar situations, such as polar expeditions with long periods of isolation and total darkness. In all exploration situations in isolated environments on Earth the explorers/crew never lost contact with their home, the Earth. Even in situations where the terrestrial environment is extremely hostile to human life, such as overwintering in polar regions with long-duration darkness and sub-zero temperatures, long-time submergence in submarines, or long-duration stays in orbital space stations, the crew can always keep contact with their home, the Earth. The astronauts in low Earth orbit have especially remarked at the view of Earth from space.

K. W. Kelley has quoted Apollo 15 astronaut Alfred Worden, as saying

FIGURE 7.5 An isolation chamber for studying the psychology of isolation. (Courtesy ESA.)

Now I know why I'm here. Not for a closer look at the moon, but to look back at our home, The Earth.

The Russian space programme has proven that a stay in low Earth orbit of 438 days onboard MIR is possible. But this evidence is based on just one cosmonaut who never experienced a period of extreme social monotony that lasted longer than a few months (due to crew exchange and visiting crews), and who got a large amount of ground-based support and teleconferencing with his relatives at home.

Missions to Mars will not be comparable to any other undertaking humans have attempted. Even though some aspects of these missions are shared by the cases described above, the physical and psychological demands made by the long journey times, the permanent living in dependence on automated life support systems, the isolation and confinement, and the lack of short-term rescue possibilities in case of emergencies will exceed anything humans have ever been exposed to.

Depending on the relative positions of Earth, Mars and the Sun, audio, video and data transmissions between ground and space will be delayed by up to 40 minutes, or may even at times be entirely blocked. Furthermore, no possibilities exist for any re-supply or short-term rescue flights. Consequently, ground-based support currently used to foster crew morale, psychological wellbeing, and mental and behavioural health of crew members during long-duration orbital spaceflight can only be minimal.

It has to be assumed, therefore, that the risks for mission success and safety associated with psychological issues that are known from different isolated and confined environments on Earth or in Earth orbit will be more severe on Mars missions. In addition, new psychological challenges that cannot be assessed in advance will arise, some of them involving risks for mental and behavioural health and mission safety.

Both transfer phases, to and from Mars, will involve long periods of comparatively low workload, monotony of environmental cues, and boredom. In addition, due to the limited size of the interplanetary parent ship, lack-of-privacy issues will be more severe than during comparable stays onboard an orbital station. Taken together, these factors will greatly increase the risk of maladaptive affective reactions and disturbances of crew performance and co-operation by interpersonal frictions and conflicts. The lack of the natural 24-hour dark/light cycle during the transfer phase will be a disturbing factor for the human circadian system, and could lead to adverse impacts on sleep quality, daytime alertness and performance which can accumulate in the course of the voyage.

Furthermore, the various highly specific skills that the crew will have gained before the start of the mission, and which will be needed on arrival at Mars, could slowly degrade because of the lack of 'on-the-job' practice. Examples of such Mars-specific skills are those related to undocking from the orbiter and landing on the surface of Mars, operating the life support systems in the Mars habitats, operating different automatic systems needed for exploration and production on Mars, and operating the experimental equipment, including tele-roboting.

To counteract these negative effects of long-lasting monotony and de-training, crew members will have to control and interact with a large variety of autonomous systems, such as information and training systems already installed on board. Besides regenerative life support systems, these systems may include robots, rovers, tele-operated devices, and, because of the lack of continuous 24-hour ground-monitoring and support typical of Earth orbital spaceflight, complex fault-management systems based on intelligent autonomous agents. Interacting with these systems will present a number of new demands that will exceed those presently demanded of astronauts in flights in Earth orbit.

On arrival at the orbit of Mars, specific psychological issues will arise from the splitting of the crew, with two crew members staying aboard the Mars orbiter for more than 500 days and the other two to four members departing for the same duration to the Martian surface. This splitting of the crew may lead to a breakdown of crew cohesiveness, since the two crew members remaining in Mars orbit may feel deprived of the kudos of participating in the tasks which are perceived as the real objectives of the mission. They may be exposed to excessive levels of monotony and boredom if they are not engaged with sufficiently demanding tasks, including monitoring and maintaining the different automated systems of the orbiter, and being the communication link between the surface crew and the Earth.

The social monotony induced by the small crew size can partially be balanced by intercom contacts with the crew on the Martian surface. If not completely engaged in the whole mission objectives, the crew in Mars orbit will have difficulties in maintaining their motivation, which will largely increase the risk of developing manifest mental and behavioural illness, including depression, anxiety disorders and psychoses. This will represent the most serious psychological limiting factor for mission success and safety.

Although to a lesser extent than the orbiting crew, the crew departing to the Mars surface will also be exposed to higher levels of monotony and boredom than those experienced in any other space or Earth-bound setting. This could adversely affect the psychological wellbeing of crew members after the first two to three months on Mars, when the initial excitement of expeditions on the Martian surface has decreased, and the tasks have become routine. Due to payload constraints, the crew habitat on the Martian surface will be much smaller than those used in other isolated and extreme environments, such as on ISS, in Antarctica and in submarines. This will considerably increase the significance of privacy issues associated with risks of maladaptive affective reactions, impairments of mood and interpersonal frictions, all of them jeopardising crew cooperation and performance.

Finally, the 'Earth out of view' aspect will involve as yet undetermined risks. Astronauts sent to Mars will be the first human beings who will effectively lose sight of the home planet. Human responses to this effect are not known and cannot be assessed in advance, but it can be assumed that the lack of any visual link to Earth will add to the feelings of isolation and loneliness. Beyond that it may induce a state of complete internal psychological decoupling from home, with as yet unknown positive or negative consequences.

On the return journeys, there will be problems of maintaining crew motivation, since the real mission, i.e. the exploration of the Martian surface, will have been completed, but the crew members will still be faced with the long journey back to Earth. This may involve elevated risks of performance degradation, and have hazardous consequences if the discipline of exercising needed to maintain conditioning for the return to Earth breaks down.

In order to counter the mission-threatening effects of psychological problems during all phases of Mars missions, the experience gained from living in space stations can be useful. In particular, the effects related to impairment of mental performance, individual wellbeing and behavioural health, and interpersonal issues can be researched.

In order to bring in preventive measures, two different types of countermeasures can be used. The first type involves basic issues of environmental engineering habitability, design of autonomous systems and work-design, including work-rest scheduling. This means adapting the living and working environment as much as possible to human needs and capabilities under the extreme conditions being discussed here. Maximum provision should also be made for privacy by providing sufficient personal space and private quarters.

The second type involves specific psychological countermeasures that might be used in order to adapt the astronaut to the living conditions in space, and to provide support during the mission. Even though the impairment of performance and psychological wellbeing of crew members or the emergence of interpersonal issues cannot be fully avoided, these countermeasures have been successful in preventing these issues from becoming a serious danger for mission success and safety. These include screening and selection of astronauts, and in-flight support of space crews. In addition, psychological pre-flight training will be included in the education programme of all astronauts.

PROTECTION OF AND FROM THE MARTIAN ENVIRONMENT

Human missions to Mars will have two interactive aspects:

i. the mission needs to be protected from the natural environmental elements that can be harmful to human health, the equipment or to their operations; and
ii. the natural environment of Mars should be protected so that it retains its value for scientific and other purposes.

The following environmental elements are threats to humans and equipment on the planetary surface:

- Cosmic ionising radiation
- Solar particle events
- Solar ultraviolet radiation
- Reduced gravity
- Thin atmosphere
- Extremes in temperatures and their fluctuations
- Surface dust

FIGURE 7.6 The Hazard Camera on NASA's Spirit rover. (Courtesy NASA.)

Although the Martian climate has some similarities with that of the Earth, e.g. four seasons, and length of the day, the environment is dramatically different from that of the Earth. Its gravity is about one-third that of the Earth and the atmospheric pressure is comparable to that on Earth at an altitude of more than 30 km. The surface radiation doses on Mars caused by cosmic rays are about 100 times higher than on Earth.

Most of these have been discussed above. The surface dust is considered as one of the dominant environmental hazards affecting robotic and human activities by contaminating moving parts, environmental control components, optical sensors, windows, human organs, etc. Terrestrial simulations of surface dust contamination have been used to further improve engineering design. Robotic *in-situ* experiments preceding the human mission can be used to verify and refine engineering design parameters related to dust contamination, and to conduct biomedical studies to determine guidelines for preventing habitat contamination.

Another component of the Martian climate is the solar UV radiation. The surface of Mars is exposed to photons of solar UV radiation of wavelengths above 200 nm. Especially the short-wavelength UV-C radiation (200–280 nm) is of high biological effect, because this wavelength range is effectively absorbed by DNA, resulting in mutations, cancer or cell death. Whereas humans can easily be protected against this harmful solar UV component, e.g. by the spacesuit or the walls and windows of the habitation module, the sensitivity of these external barrier materials to UV radiation needs to be determined. Using greenhouses for bioregenerative life support purposes requires also the selection of suitable windows with appropriate cut-off features: the plants might require a certain fraction of UV, especially in the UV-A and UV-B regime, whereas the short-wavelength part of UV-B and the UV-C radiation will certainly be harmful for their growth.

In order to assess the impact of solar UV radiation on putative ecosystems on Mars, some information can be gained from studies simulating the UV radiation climate of Mars, e.g. in Mars simulation chambers or onboard the ISS using appropriate optical filtering systems. A first step in this direction will be done in the EXPOSE facility to be mounted on the external part of the International Space Station, as has been described by G. Horneck and others in 1999. It is also necessary to establish guidelines for human exposure to the solar UV spectrum, as well as for cultures used for bioregenerative life support.

The second major part is to protect the pristine environment of Mars from harmful influences caused by the human invasion. The risks of contamination involved in the presence of humans on Mars are threefold:

i. the risks to the crew from Martian microbes (if any exist);
ii. the risk to life on Earth via returned Martian samples (accidentally or deliberately brought back to Earth) and;
iii. the risk to Mars from imported terrestrial micro-organisms.

The planetary protection protocols accepted by the international community and recommended by the Committee of Space Research (COSPAR) require in the case of Mars the following measures:

- All vehicles landing on Mars and their equipment should be assembled in a clean room.
- They should be sterilised by appropriate procedures to reduce the microbial load to the level accepted for the first automatic landers on Mars, i.e. the famous Viking missions which in 1976 searched for the first time for signatures of microbial life in the Martian regolith.
- Recontamination should be prevented by enclosing the sterilised lander with its equipment in a bioshield, as has been described by D. L. DeVincenzi in 1992.

These requirements reflect the need to prevent contamination of the Martian surface with terrestrial micro-organisms – which would jeopardise the chance to detect life forms indigenous to Mars – and thereby preserving the integrity of any 'search for signatures-of-life experiments'.

However, these requirements can only be met with robotic missions to Mars. The scenario changes when humans are involved in the mission. Since humans naturally carry vast amounts of microbes required to sustain important body functions, Mars will inevitably become contaminated with terrestrial micro-organisms as soon as humans arrive on its surface. Although the surface of Mars in general seems to be very hostile to microbial life, it cannot be excluded that some micro-organisms accidentally imported may find protective ecological niches where they could survive or even metabolise in, grow, and eventually propagate. Having this in mind, the ESA Exobiology Team stressed the vital importance of a substantial series of robotic missions to precede the human mission to Mars in order to carry out the essential exploratory search for life by *in-situ* measurements at selected sites, as has been described by A. Brack and others in 1999.

As a consequence of human activities on the surface of Mars, the planetary environment may change with time. This may become especially severe if natural resources are used on a large scale.

FIGURE 7.7 The Franco-Italian Concordia base under construction in Antarctica. It will be used by ESA for studies of long-term space missions. (Courtesy ESA.)

Therefore, a systematic recording of the environment needs to be made, starting with the first human landing. The natural release of gases may provide information about the interior structure and may also be of use for establishing a permanent human base. Such natural gas releases may be detectable now, but could become difficult to detect with more intense exploration. Therefore, a search for natural transient gas releases should be made from orbiters or by the use of surface sensors before the establishment of a permanent human settlement. Furthermore, it is very unlikely that all environmental effects on the Martian environment from human activities on its surface can be predicted. Therefore, records of human activities on the planetary surface should be maintained in sufficient detail so that future scientific research can determine whether environmental modifications have resulted from such human activities.

8 Human exploration and the search for life

They (remote observations) emphasise topography or chemistry but they cannot define the relationships among the rocks. Geologists touch rocks, peer at them, push them, even smell them – to find out what is there. And while they are doing this, they constantly think and rethink. They quickly change from observing the horizon to studying a sand grain as they do so. They rapidly manipulate tools and samples. A robot is not only slow and unproductive, it is incapable of keeping up with human cognition.

Graham Ryder (1996)

Human researchers are endowed with capabilities which it is difficult to see will, in any near- to medium-term future, be acquired by machines. These are chiefly flexibility and the ability to adapt and improvise, as well as the ability to exercise judgement based on experience and professional training. The non-quantifiable quality of intuition and the role of serendipity that can be rapidly exploited by humans on the spot can hardly be matched by machines, however advanced. Without denying the role to be played by *in-situ* analysis by robots and even robotic sample return missions, it is clear that human exploration will bring a quantum leap in capabilities for the search for life and, beyond that, the exploration of the ecologies which will be revealed after the initial discovery itself.

How much experience to date do we have about the capabilities of men and women acting both independently and cooperatively in extreme environments to cope, to create incremental opportunities and results, and finally to thrive in hostile conditions in the pursuit of scientific advancement and exploration? We will look at five cases where humans are or have been put to the test as intelligent and artful beings, working and solving problems in hostile environments:

- Extravehicular activities as a whole, starting with the NASA and USSR spacewalks of the 1960s.
- Early and more recent space station activities, such as on Skylab (USA), Salyut (USSR), more recently the Russian MIR station, and the presently on-going assembly and construction activities of the International Space Station (ISS).
- The Apollo Moon missions.
- The Hubble Space Telescope repair and maintenance missions.
- Terrestrial analogues, such as deep-sea operations, extreme climate exploration and Antarctic base operations.

Spacewalks of the 1960s

Soviet cosmonaut Alexei Leonov became the first person to walk in space in March 1965. He wriggled out of his spacecraft Voskhod II and stayed outside for a mere 12 minutes before re-entering. In the process, he became stuck in the inflatable airlock and nearly perished due to exposure to the Sun before wrenching himself free and closing the hatch. The dangers inherent in spacewalking became evident.

In June of the same year, American astronaut Ed White left his Gemini 4 craft and floated in space, attached only by a 25-foot tether and umbilical. After White's flight, NASA engaged in a bold programme of spacewalks using the Gemini flights. But it was not all plain sailing. When Gemini 9 astronaut Gene Cernan tried to work in space, he found the microgravity environment too disorienting while outside the spacecraft. He became overheated and exhausted and had to abandon the spacewalk. But as time progressed and more missions and spacewalks were made, the Soviets and NASA learnt more about how to conduct these difficult forays. In January 1969, two Soviet cosmonauts transferred from one Soyuz capsule to another and returned to Earth. NASA developed a system of handholds and restraints that greatly aided working outside the spacecraft. The Apollo spacesuit that was designed for moonwalking provided much better cooling of the astronauts and greatly improved ease of movement.

FIGURE 8.1 Ed White becomes the first human to float freely in space. (Courtesy NASA.)

SKYLAB

After the termination of the Apollo programme in the early 1970s, and with a strongly reduced budget, NASA decided on a more modest space station programme called Skylab. Using the Saturn V technology developed for Apollo, the third stage would be used in a 'dry' mode, outfitted as a laboratory for experiments in low Earth orbit. Astronauts would be launched to Skylab and return to Earth in Apollo spacecraft.

On the first launch of Skylab on 14 May 1973 and 60 seconds after lift-off, the meteorite shield designed to protect Skylab was accidentally deployed and ripped from the ascending spacecraft complex by the drag of the atmosphere. As it tore away, it damaged both of the solar panels. When it reached orbit, Skylab was on reduced power and was badly overheating. Engineers on the ground spent the next 10 days getting it into a stable and safe configuration. By then it was clear that direct human action in the form of astronaut intervention would be needed.

Three three-man crews were successively launched to Skylab on 25 May, 28 July and 16 November 1973. During a crucial EVA, the first crew deployed a parasol-type sunshield through the solar scientific airlock and later released a new solar array. Again, during another EVA the second crew managed to erect a second sunshield. The actions of these crews saved the Skylab as a viable laboratory in space. The NASA conclusion on the ability of the crew to handle complex repair and advanced tasks declared:

> *The effectiveness of Skylab crews exceeded expectations, especially in their ability to perform complex repair tasks. They demonstrated excellent mobility, both internal and external to the space station, showing man to be a positive asset in conducting research from space.*

It went on to talk about the ability of humans to do research in space:

> *By selecting and photographing targets of opportunity on the Sun, and by evaluating weather conditions on the Earth and by recommending Earth Resources opportunities, crewmen were instrumental in attaining extremely high quality solar and Earth oriented data.*

By the mid 1980s, EVA had become a fairly routine affair in most human spaceflight missions.

THE CASE OF THE HUBBLE SPACE TELESCOPE

After nearly 30 years of preparation, planning and development, the Hubble Space Telescope was launched with the Space Shuttle Discovery as mission STS-31 from Launch Complex 39B of the Kennedy Space Center on 29 April 1990. Hubble had been touted as the wonder-instrument likely to lead to some of the greatest advances in observational astronomy in a long time. The deployment of the telescope was not straightforward and there were problems with the solar panels. At that stage, NASA had already warned two astronauts to get ready for EVA to solve the solar panel problem. Ten minutes short of them having to start the EVA, the problem was solved by a software fix. There were also pointing problems and the telescope often went to curl-up in a safe mode. But all this was to be expected from one of the most sophisticated instruments in the civilian space field. There followed weeks with the engineers working to ready the telescope for observations until finally NASA was forced to admit, to the dismay and anger of the scientists, that the primary mirror of the telescope suffered from spherical aberration: the primary mirror was wrongly shaped. Instead of sharp images of stars and other celestial objects, the images were fuzzy. For NASA it was a public relations disaster.

In early December 1993, Shuttle mission STS-61 powered into orbit with a highly trained crew determined to 'fix' Hubble. Over a period of nearly a week, the world watched as the crew accomplished an almost complete overhaul of Hubble – inserting correcting optics to compensate for the spherical aberration of the primary mirror, replacing gyroscopes, solar panels and memory units. During this remarkable work the crew dealt splendidly with jammed instrument bay doors and flying nuts and bolts. The advocates of manned spaceflight had a field day. Everyone has by now seen some of the vast array of amazing, crisp new images of the plethora of celestial objects that Hubble has imaged since then.

The lesson was clear. Sophisticated science/technology hardware needs human supervision and intervention.

THE APOLLO PROGRAMME

The Apollo programme has been humanity's only foray into human exploration of other planetary bodies. After President Kennedy's decision to send men to the Moon and return them safely to Earth in 1961,

FIGURE 8.2 The Shuttle astronauts repair the defective Hubble Space Telescope. (Courtesy NASA.)

Neil Armstrong became the first human to take the small step and the great leap onto the lunar surface on 21 July 1969. Seven other missions to the Moon in the Apollo series took place between 1969 and 1973 when the programme was abandoned. Lunar samples weighing 380 kg were returned to Earth and a relatively extensive survey of the lunar surface was carried out, notably employing mobile 'buggy' vehicles for the more extensive forays of about a kilometre or so from the landing site. The last mission (Apollo 17) took trained geologist Harrison Schmitt, who carried out the first real geological fieldwork.

We will not make a detailed assessment of the Apollo missions here as they have been extensively analysed and described elsewhere. In particular, the excellent *A Man on the Moon: The Voyages of the Apollo Astronauts* by Andrew Chaikin (1994) covers all this brilliantly, especially the advantages of having trained geologists and, implicitly, in the search for life, trained biologists/exobiologists active on the planetary surfaces being investigated.

However we can note that the Apollo programme demonstrated the following:

- Setting a high-level national or international goal for human-based exploration in planetary exploration can succeed.
- Significant human intervention capability at all stages is mandatory. Neil Armstrong's last-minute piloting of the Lunar Lander of Apollo 11 to avoid a crash in a crater, and the critical operational decisions needed to return the Apollo 13 astronauts safely to the Earth after a catastrophic failure are examples.
- Mobility is a paramount requirement, and the more mobility away from the landing site/base the more science and exploration pay-offs there will be.
- Science is not necessarily a driver in the programmatic or mission decisions but can be a major beneficiary if there is an intelligent use of the opportunities opened up, as has been argued by I. A. Crawford.
- Contrary to the claims of critics, the technology pull of devising ways to ensure humans can enter and survive the extremes of space environments can have enormous pay-offs in wide swathes of technology development such as:
  - advanced life support systems
  - biomedical monitoring remotely
  - advanced propulsion

THE SALYUT AND MIR SPACE STATIONS

Since April 1961, and the first orbital flight by Yuri Gagarin around our globe, the Soviet Union pursued an aggressive programme of human spaceflight concentrated on capsules and space stations in low Earth orbit. By the early 1960s, the United States, under Kennedy's administration, and with the successes of the Mercury (1 man), the Gemini (2 man) low Earth orbit capsule missions behind it, had set its sights on the Moon with the Apollo programme. In a sense, this divergence of approaches reflected the differing political and cultural realities in each country.

Whereas the American public tired quickly of low Earth orbit space shots and quickly embraced the challenge of sending humans to the Moon and returning them safely to Earth, this in turn proved to be an ephemeral challenge which, once achieved, was quickly abandoned in the 1970s after Apollo 17. Between the first landing on the Moon – Apollo 11 in July 1969 and including the Apollo 13, which had been a near-fatal failed landing – and the last landing, Apollo 17 in 1973, a mere six missions constituted the whole of the vastly expensive Apollo programme.

The Soviet programme reflected a less public-relations-oriented approach and also a more centralised, longer term type of planning in the classical Soviet style. It focused on the building up of a low Earth orbit space-station infrastructure, with the Salyut 2-man space station orbiting the Earth in one form or another from the mid 1970s into 1985, when the MIR 3-man space station was launched. By any accounts, the Salyut and MIR programmes must be considered as enormous achievements, opening the way for a more profound appraisal of the ability of humans to deal with the long-term exposure to space conditions, notably zero-gravity and isolation, both physiological and psychological.

In a sense, the Russian experience, although sometimes lacking a certain systematic scientific approach, represented a dogged long-term attack on these problems. Characteristically, the USA pursued more upfront, in the public view, space activities. The Space Shuttle, designed in the 1970s and originally touted as a low cost, quick turnaround, multipurpose launch vehicle, never achieved such a status and remains to this day a high profile, low frequency launch system, with each Shuttle launch featuring new public relations and publicity angles. Tragically, in 1986, the first launch of a high-school teacher – another publicity event – ended in the Challenger disaster and immediately afterwards the decision of the US Department of Defense to abandon the Space Shuttle as a reliable vehicle for delivering sensitive military spy-satellites to orbit. The traumatic loss of the Columbia in February 2003 compounded the image of the Space Shuttle as a poorly conceived space transportation system.

The MIR space station ended its life in a planned destructive burn-up over the Pacific Ocean in 2001. It had been in orbit and housed a crew of three on a continuous basis for 16 years. In the end, after the fall of the Soviet empire and the *rapprochement* between the USA and Russia in space affairs, it had been adapted to rendezvous with the NASA Space Shuttle and had seen the visit of a number of US and European astronauts over the final years of its life. This involved a significant if not bright-lights programme of scientific research in life sciences and physical sciences, as well as research in astronomy and Earth observation. The end of MIR also saw the occurrence of a number of potentially catastrophic events, including a fire and a collision which induced a partial depressurisation. These events have given valuable data on the operational contingencies which space stations or other human-bearing spacecraft must be adapted to. As an example of international cooperation in space exploration and the pursuit of our understanding of the effects of the long-term space exposure of humans to space conditions, MIR has had, up to now, no equal.

THE INTERNATIONAL SPACE STATION (ISS)

Never has an international science and technology project been so large and so replete with anguish. Conceived in the 1970s by NASA as the logical destination of the 40–50 Space Shuttle flights that NASA would carry out every year, the idea remained in limbo, as the shrunken NASA human spaceflight budget went largely into the development of the Space Shuttle. In 1984, President Ronald Reagan, as part of his opposition to the Soviet Union, thought that an international science and technology project in space would be a good way to demonstrate the resilience of the 'free' world to the challenges of the Soviet empire. Space Station Freedom was born involving the USA, Japan, Canada and most of the member states of the European Space Agency. With an estimated development cost of about $8 billion in

1984 dollars, and a projected operational cost of about $20 billion over 10 years of life, it would be the largest international science and technology project every undertaken.

By 1993 and having consumed many billions of the 1984 dollars in studies, Space Station Freedom was essentially dead. The Clinton administration, tired of all the wrangling and keen to get the Russian Space Agency involved as a means to ensure that Russia would not sell rocket technology to countries like Iran, sent this Space Station into another orbit. After endless re-configuration assessments, it ended up with significant Russian involvement in on-orbit infrastructure and the use of Soyuz and Progress launchers for re-supply and crew transportation. It is now known as the International Space Station (ISS), and despite the recent Columbia accident, is now nearing completion. Present estimates of its total development and operational costs to so-called 'Assembly Complete' (about 2006) put these at around $100 billion. The Space Station will have a permanent crew of seven when fully operational, operate over a period of 20 years in low Earth orbit, and help to answer many questions about the long-term survival of humans and robots in space. It may well eventually serve as a staging post for the assembly of large infrastructure elements in space, function as a quarantine post for sample return missions to Mars and, with the development of an crew transfer vehicle dedicated to the sole task of transporting humans from Earth to low Earth orbit, may serve as a collecting and rendezvous point for humans in space. The ISS in its final state could be capable of accommodating a human crew complement of up to 10–15 persons.

Once the idea is accepted that the Space Shuttle was an ill-conceived vehicle that tried to combine both human spaceflight and the delivery of cargo to low Earth orbit in a potentially dangerous way, the political decisions to escape this situation will be made. As we will see, the approach of Robert Zubrin foresees this. One must bring the infrastructure separately to the point of rendezvous in lower cost non-human-rated vehicles. Then one brings the humans separately in the more reliable but more expensive human-rated vehicles. It all seems so obvious now.

ANTARCTIC BASES

Antarctic bases are increasingly seen by space agencies as almost ideal test grounds for simulating the potential of humans to survive, in often closed communities, in isolation for relatively long periods of time (6–12 months or even more). Studies now being initiated in joint cooperation between the European Space Agency and the national Antarctic agencies of France and Italy, using the Franco/Italian Concordia Station now under construction, will attempt to gauge how well these kinds of episodically isolated Antarctic stations could simulate, for example, isolated outposts on Mars. Questions being asked are:

- Can the effects of isolation for 7–9 month winter periods in Antarctic stations simulate similar periods on the transit to and from Mars in relatively small spacecraft?
- What about the possible 500-day stay on Mars in some scenarios? Can we learn something from Antarctic bases about such long stays?
- What really are the physiological, psychological and psycho-physiological effects on teams of humans subjected to sometimes extreme and isolated conditions, and can they be modelled by Antarctic base simulations?

Finally, space agencies are looking at other confinement/isolation scenarios in which teams of people/crew must work together in a coordinated fashion over long periods of isolation and potentially extreme conditions. Chief among these, of course, would be the experience of crews of nuclear submarines. Experience from this somewhat secretive area of military operations will also be relevant here.

HUMANS VERSUS ROBOTS: A FALSE COMPARISON?

Since the earliest days of human spaceflight, an intense and sometimes rancorous debate has pitted the proponents of human spaceflight against those who argue that almost everything can be done in space using robotic means. The argument was at its height in the years leading up to and during the Apollo missions to the Moon. It still goes on today and will certainly reach new heights of emotion when the decisions for a real commitment to human missions to Mars are in prospect. The Apollo furore was led by a significant section of the scientific community, chiefly astronomers, who argued that the exploration of the Moon could as easily be achieved by robots as by the vastly more expensive human missions. They argued that the money saved could be used to do other science such as, perhaps inevitably given their perspective, astronomy.

However, British scientist I. A. Crawford and others have argued that, contrarily, the supposedly saved money would probably be used by politicians for purposes other than science and that human spaceflight itself brings fresh money to do the science that can be done by humans in space, such as, for example, the astronomy observatories that are presently planned to be taken to the human-crewed International Space Station.

In addition, given the example cited above, it should be clear that humans can be extremely effective at complex tasks requiring dexterity, curiosity in problem solving, and sophisticated repair and maintenance activities. It would be impossible for robotic technology available today to be able to carry out the complex repairs to Skylab and the Hubble Space Telescope described above. To approach the issue more subtly, we need to highlight the complementary capabilities of both humans and robots. We ask:

What are robots good at?

- repetitive tasks
- pre-programmed operation
- continous operation if enough power is available
- operation in hostile or extreme environments
- boring tasks
- pre-programmed multitasking
- parallel complex analysis (e.g. of soil samples)

What are humans good at (or at least better at than robots)?

- decision-making/management
- adaptive behaviour

- informed guessing
- operation in low information/confusing situations
- communication with other humans
- organising new tasks
- identifying new opportunities or threats
- non pre-programmed multitasking
- re-planning/re-prioritisation
- integrating unexpected information

There would seem to be really no contest here. The case has perhaps been overstated regarding the limitations of robots and how versatile humans can be, and recent developments in the area of artificial intelligence may modify all this somewhat; but by and large it seems difficult to escape the conclusion that for the foreseeable future, in the context of space exploration, robots will play an essentially complementary, automatic support role to humans.

9 Interplanetary ethics

Any human search for life on an extraterrestrial body will necessarily occur in three stages:

- the *in-situ* robotic search for life
- sample return to Earth for a more accurate analysis
- human exploration

Each stage leads to several ethical and philosophical questions that are developed in this chapter.

WHAT IS 'ETHICS'?

Historically, the basic question of ethics has been posed in two main ways. Socrates and the Greek philosophers asked: 'How should we live?', while the moderns ask with Kant: 'What should we do?' The Greek (*ethos* and *ethicos*) and Latin (*mos* and *moralis*) etymologies lead broadly to these two approaches. The first sees ethics as a set of rules agreed by the members of a society and expressed in terms of notions of good and evil. The second one considers ethics as a philosophical discipline that analyses these rules by exploring their foundations and purposes. Some American writers use the expression 'Ethics and Values Studies' to stress the philosophical dimension of ethics. Apart from philosophy, however, ethics and morals have also constantly to cope with the practical aspects of human existence, so that as well as the generally expressed obligations and duties, human beings must also be able to rely on their own judgement and follow their conscience according to criteria other than just the strict obedience to law.

WHY SEARCH FOR EXTRATERRESTRIAL LIFE?

Gazing at the stars, humans have always wondered if there are other beings out there. The idea that life could be universal can be traced back to the Greek philosophers. In 300 BC, Epicurus wrote to Herodotus: 'There is an infinite number of worlds ... There is no reason why these worlds could not contain germs of plants and animals and all the rest of what can be seen on Earth.' This age-old dream was further formulated by the Roman philosopher Lucretius: 'Confess you must that other worlds exist in other regions of the sky, and different tribes of men, kinds of wild beasts'. For agreeing with Lucretius, the defrocked Dominican friar Giordano Bruno was burned at the stake in 1600 on Piazza Campo dei Fiori in Rome, by order of the Inquisition. With time, the idea became less and less scandalous (and dangerous to its believers), and in 1686 the French poet and philosopher Bernard de Fontenelle published his *Entretiens sur la pluralité des mondes*.

Along with this human curiosity about other beings on other worlds, there is a scientific motivation to search for a second genesis of life. As already mentioned, life – defined as a chemical system capable of self-reproduction and of evolution – originated from the reaction of reduced carbon-based organic matter in liquid water. Schematically, the prebiotic resources needed for the emergence of life can be compared to the elements of a chemical robot. By chance, some parts assembled to form a robot capable of assembling other parts to form a second identical robot. From time to time, a minor error in the assembly process generated more efficient robots. The number of parts required for those first robots is still not known. The problem is that, on Earth, those earliest parts have been erased by geological processes such as plate tectonics, and by life itself.

The chances of understanding the emergence of life on Earth by recreating a similar process in a test tube will obviously depend upon the simplicity of the chemical reactions that lead to life. The discovery of a second genesis of life on another celestial body would strongly support the idea of a rather simple genesis of terrestrial life and may perhaps, with luck, provide information about those early chemical robots.

CAN WE TAKE THE RISK OF CONTAMINATING EXTRATERRESTRIAL BIOLOGICAL HABITATS?

The unintentional release of terrestrial micro-organisms on Mars is possible as soon as a spacecraft touches the Martian surface with terrestrial micro-organisms that are inadvertently transported to Mars. The biological contamination of Mars would be particularly serious during the present period when we are trying to search for life and prebiotic chemistry on the red planet, particularly because micro-organisms have been shown to be extremely resistant. For example, on 20 April 1967, the unmanned lunar lander Surveyor 3 touched down near Oceanus Procellarum on the Moon. Two-and-a-half years later, on 20 November 1969, Apollo 12 astronauts Charles Conrad and Alan Bean recovered the Surveyor 3 camera. When NASA scientists examined the camera back on Earth, they were surprised to find specimens of *Streptococcus mitis* that were still alive. These bacteria had survived for 31 months in the vacuum, dryness and temperature extremes of the Moon's surface although contamination by the astronauts cannot be totally ruled out.

In order to study the survival of resistant microbial forms under harsh conditions, microbial samples have been exposed *in situ* aboard balloons, rockets and spacecraft – such as Gemini, Apollo, Spacelab, the Long-Duration Exposure Facility (LDEF), FOTON and EURECA – and their responses investigated after recovery. All showed themselves to be extremely resistant (see Chapter 4).

Even if most of the bacteria brought to an inhospitable planetary surface do die, some may find sheltered ecological niches that allow them to survive, to propagate and eventually to mutate. Whether dead or alive, the imported microbes will progressively contaminate the planetary environment, which when viewed from an exobiology viewpoint poses problems of identifying which microbes are really native to that body and which are not.

Planetary protection issues are covered by international United Nations treaties drawn up in 1967 and 1979 on principles governing the activities of states in the exploration and use of outer space, including the Moon and other celestial bodies. The intent is twofold: (1) to protect the planet being explored from contamination by terrestrial microbes and organics in order to maintain its pristine state, and thereby safeguarding the integrity of exobiology studies in the search for evidence of an indigenous micro-organism, including precursors or remnants of life (i.e. forward contamination); and (2) to protect the Earth from the potential hazards posed by extraterrestrial matter carried on a spacecraft returning from another celestial body (i.e. backward contamination). Guidelines have been elaborated for specific space-mission/target-planet combinations, such as orbiters, landers, or sample return missions, and are laid down in the recommendations of the Committee on Space Research (COSPAR) of the International Council of Scientific Unions. This has all been described by D. L. DeVincenzi and P. D. Stabekis in 1984; J. D. Rummel in 1989; and J. D. Rummel and others in 2002. An updated COSPAR Planetary Protection Policy was approved on 20 October 2002 by the Bureau of COSPAR at the World Space Congress in Houston, Texas. The new policy defines five categories for target-body/mission-type combinations and their respective suggested ranges of requirements. Category I includes any mission to a target body which is not of direct interest for understanding the process of chemical evolution or the origin of life. No protection of such bodies is warranted and no planetary protection requirements are imposed by this policy. Category II missions comprise all types of missions to those target bodies where there is significant interest relative to the process of chemical evolution and the origin of life, but where there is only a remote chance that contamination carried by a spacecraft could jeopardise future exploration. The requirements are for simple documentation only. Category III missions comprise certain types of missions (mostly fly-by and orbiter) to a target body of chemical evolution and/or origin of life interest, or for which scientific opinion indicates a significant chance of contamination which could jeopardise a future biological experiment. Requirements will consist of documentation and some implementing procedures, including trajectory biasing, the use of clean rooms during spacecraft reflected by the absolute prohibition of destructive impact upon return, the need for containment throughout the return phase of all returned hardware which directly contacted the target body or unsterilised material from the body, and the need for containment of any unsterilised sample collected and returned to Earth. Post-mission, there is a need to conduct timely analyses of the unsterilised samples collected and returned to Earth, under strict containment, and using the most sensitive techniques.

If any sign of the existence of a non-terrestrial replicating entity is found, the returned sample must remain contained unless treated by an effective sterilising procedure. Category V concerns are reflected in requirements that encompass those of Category IV, plus a continuing monitoring of project activities, studies and research (i.e. in sterilisation procedures and containment techniques).

TO WHOM WILL EXTRATERRESTRIAL MICRO-ORGANISMS BELONG, IF ANY?

On Earth, extremophiles are sought by biochemical and drug companies for their special properties. For example, the Polymerase Chain Reaction intensively used in biochemistry and biology to amplify DNA sequences has been made possible thanks to DNA polymerase T7 extracted from thermophiles. Bacteria capable of surviving temperatures below the freezing point of water are of interest for the cryogenics industry, which has already patented two bacteria. The property of putative extraterrestrial bacteria must therefore be defined legally. Could they be patented? *Res communis* or common property of humanity? These questions remain open.

THE ETHICS OF HUMANS IN SPACE

The presence of humans in space raises the questions of 'why' and 'how', both being associated with ethical concerns. It is often argued in discussions on humans in space that: 'the destiny of humanity is to occupy space, a destiny written in our genes'. But opponents can still argue that the Earth can be seen as a spaceship driven by humanity acting as a crew, and it is the destiny of a crew to stay onboard the ship for which it is responsible. But what is the real destiny for humanity? Lost in tens of billions of galaxies, made of the same biological matter as tens of billions of other living species, is a human being more than just a single speck within a gigantic blind mechanism? So far, humanity has conquered almost all possible places on Earth and is about to explore and conquer space. In so doing, humanity has progressively moved from a 'natural world' of contingency and constraint towards a more 'cultural world' of greater choices. In this perspective, the presence of humans in space is not driven by some genetic imprint or evolutionary determinism. Rather it is derived from humanistic motivations arising from a certain cultural and historical background.

Arguments in favour of the presence of humans in space can sometimes be financial, based on the cost of manned missions compared to the corresponding unmanned robotic missions (repairing spacecraft in Earth orbit, for example).

The contribution of human expertise must also be considered. Concerning the exobiology exploration of extraterrestrial celestial bodies, such as Mars and Europa, exobiology research will be done, at first, by *in-situ* measurements using robotic systems, probably followed by sample return missions. However, ultimately, the intervention of humans is needed, just as it is on Earth in a research environment, to do those tasks that cannot readily be done by robots. These tasks primarily involve the exercise of judgement, based upon extensive professional and personal experience, coupled with flexibility and ingenuity to adapt and to improvise in real time. In order to fully benefit exobiology research, and its associated fields of mineralogy and geochemistry, it is therefore mandatory that some of the crew have professional scientific training in these fields. Without that expertise, the value of their presence will be substantially diminished, just as it would be in a terrestrial laboratory. Although robotic exploration attracts huge public interest (the Mars Pathfinder website was visited more than one billion times in one month), only humans in space can offer to the person in the street the possibility to identify themselves with astronauts, as happened during the Apollo missions exploring the Moon.

As for the 'how', the sending of humans to space must fulfil certain ethical rules which respect the crew, their security and integrity. The risks that will be involved in the human presence on extraterrestrial bodies relate mainly to the risk to the crew from extraterrestrial microbes, if any, being brought into the base by the crew after an outdoor exploration. Since the base provides the crew with adequate temperature and humidity conditions, it could allow the growth of putative dormant cells. This applies to professional astronauts as well as to potential tourists in space.

The flight of Dennis Tito in May 2001 to the ISS opened the era of space tourism, but at a relatively high cost and environmental impact. The cost, 20 million dollars, is rather excessive, and in the absence of any fall in such costs can hardly lead to a more popular tourism. On the other hand, low Earth orbit is crowded with debris and must already be protected against any wild space tourism. Clearly ethical guidelines also will need to be developed in this area.

TERRAFORMING A CELESTIAL BODY: EXPLORATION OR CONQUEST?

Terraforming consists of altering a planetary surface and its atmosphere to suit human needs. In a famous paper published in *Nature*, Christopher McKay and others (see reference in Chapter 4) considered that our introduction of greenhouse gases into the Earth's atmosphere at rates sufficient to modify the climate has led to suggestions that we should take active global measures to counteract the predicted global warming, in effect 'terraforming' the Earth.

According to these authors, Mars is believed to be lifeless today, but it may be possible to transform it into a planet suitable for habitation by plants, and conceivably humans. If it is feasible, should we consider terraforming Mars? This raises ethical and moral questions. Does Mars, as a planet, have an intrinsic value in and of itself? Do we have the right to colonise Mars, use the resources that are available there, or should we leave them as they are? Robert Haynes and Christopher McKay have come up with a thorough list of pros and cons. Among the arguments in favour of terraforming, it is argued that such an activity would provide a long-term project on which humans could focus, with a goal that was both useful and desirable for humans. Also, an active biosphere on Mars would provide a refuge for life on another planet in the Solar System in the event of war or natural global catastrophe that might destroy all life on Earth.

Some arguments against terraforming Mars include the fact that it is impossible to prove conclusively that Mars is totally devoid of life today, and that the project might extinguish an indigenous Martian life form. Also, it is said that since humans have done such a bad job of managing the Earth's environment it is presumptuous to imagine that they can be wise and successful planetary engineers of Mars.

EXTRATERRESTRIAL HABITATS, '*RES COMMUNIS*' OR PATRIMONY OF HUMANITY?

When determining the legal status of extraterrestrial territories, a distinction should be made between two concepts:

- The territory belongs to nobody and is therefore accessible and exploitable by everybody. This status was recognised as '*res communis*' by Roman law.
- The territory belongs to everybody and must, therefore, be managed for the common interest, with an equal access to the resources. To this category belongs what is generally designated as 'common patrimony of humanity'.

The distinction between the two regimes is not always easy to formalise but there is, at least, one example on Earth where it is operational. The ocean floor beyond the territorial waters of states, including the exploitation of the metallic nodules, has been declared as being part of the common patrimony of humanity. The ocean waters above the floor, i.e. the open sea, have the status of *res communis*: their access must be free and no state can claim any kind of sovereignty. This regulation has been operating since 1994 and even if it has not been ratified by all states, it is apparently acknowledged by everybody (probably because the exploitation of the ocean floor for metallic nodules is far from being an easy task).

What about space? The treaty signed in 1967 specifies: '§1. The exploration and the use of extra-atmospheric space, including the Moon and other celestial bodies, must be done for the good and the benefit of all the countries, irrespective of their degree of economical and scientific development. §2. The extra-atmospheric space, including the Moon and other celestial bodies, can be explored and used freely by all states without any discrimination, under egalitarian conditions and in conformity with international law, all regions of celestial bodies being kept accessible freely.' This treaty is ambiguous. The first article seems to favour the status of common patrimony, but the second article weakens the point by suggesting a status of *res communis* since the states have no obligation to impose special conditions on their nationals concerning the exploitation of the riches. In brief, the situation is similar to the regulations governing the oceans, probably because the commercial exploitation of the resources of space, the Moon and other bodies are still very limited.

TOWARD A SPACE HUMANISM?

As we view the future of space exploration and exploitation from a humanist perspective, we can see that space sciences and technology, like life sciences and biology research today, will offer ways to deepen the development of human potentialities. Space exploration and the technologies it spawns will unleash new, creative forces of knowledge, reason and genius. This humanist aspect of space exploration cannot be static, and humans will have to constantly adapt their choices by integrating the meaning of new discoveries and adopt behaviours somewhere between sterile conservatism on the one hand and the reckless pursuit of hasty exploration and exploitation on the other. This opportunity for strengthening humanism will link the notions of choice, responsibility and freedom. Space exploration will bring the possibility to explore the other planets of the Solar System and to scrutinise the Universe, thus enlarging our cosmic world-view. It will give us a way of looking at our own planet from the outside, at its clearly finite resources and at the all too evident and sometimes deleterious effects of human activities.

PLATE 1 Geologist-astronaut Harrison Schmitt doing fieldwork during Apollo 17. (Courtesy NASA.)

PLATE 2 An astronaut-bearing Soyuz capsule prepares to dock at the International Space Station. (Courtesy ESA.)

The images in this plate section are available in colour as a download from www.cambridge.org/9780521124546

PLATE 3 Water, light and life. (Photo V. Rouse-Delimol.)

PLATE 4 Comet Hale-Bopp and a meteorite trail. The Earth is bombarded by cosmic material on a constant basis. (Photo James W. Young – Resident astronomer JPL/NASA/Table Mountain – used with permission.)

PLATE 5 Unmelted micro-meteorites (black grains) from old Arctic blue ice. (Courtesy ESA.)

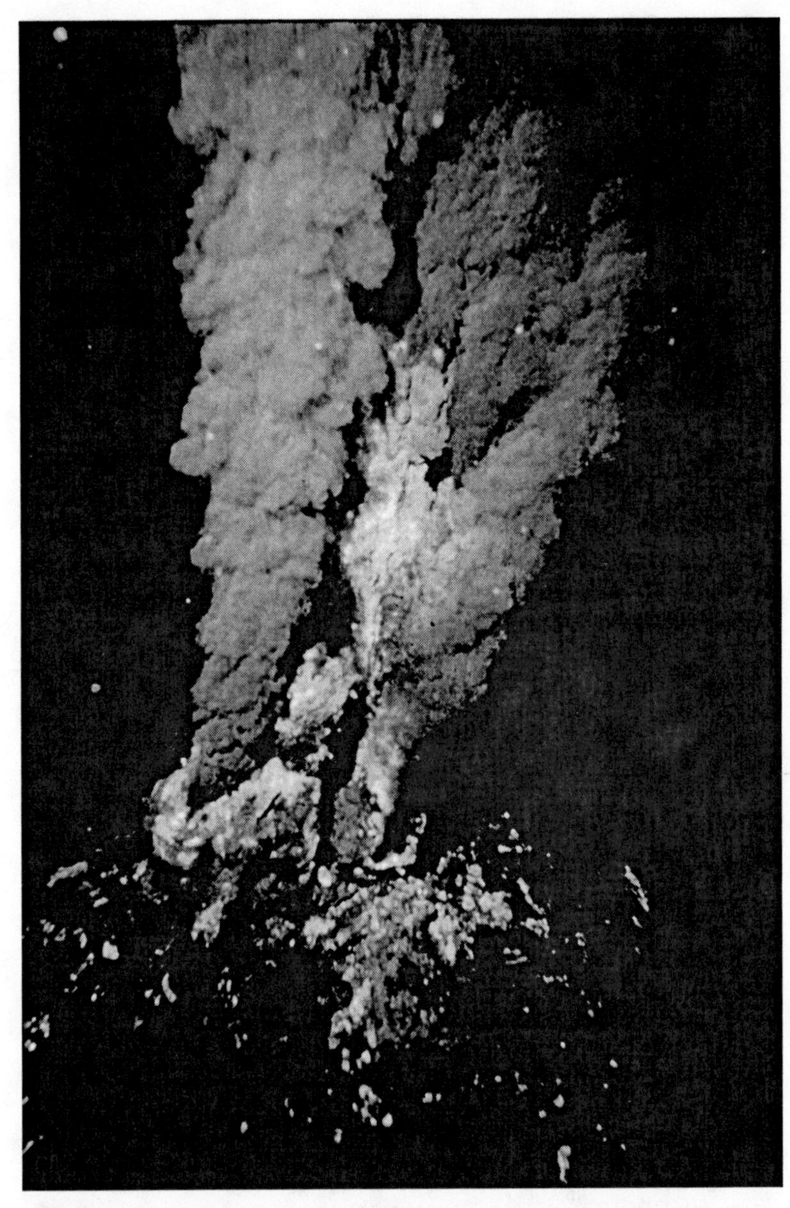

PLATE 6 A hydrothermal vent and black smoker. (Courtesy NOAA.)

PLATE 7 Evaporite with horizontal bands of endoevaporitic microbial communities (From ESA SP-1231).

PLATE 8 Typical microbialite (or 'stromatolite') from the Precambrian Transvaal Dolomite about 2.3 billon years old. The bun-shaped and partially interfering layers represent consecutive growth stages of the primary microbial community. (From ESA SP-1231.)

PLATE 9 Evidence for the pervasiveness of intense radiation in interplanetary space can be seen in spectacular auroras. These are created in the upper atmosphere by the interaction of the solar wind and the intense magnetic field of a planet. Here the Galileo mission captures Jupiter's aurora (above) and the Hubble Space Telescope shows auroras in Saturn's northern and southern hemispheres (below). (Courtesy NASA/ESA.)

PLATE 10

PLATE 11 Artist's impression of how astronauts will work on the platform where ESA's EXPOSE facility for the exposure of biological materials to the space environment will be located. (Courtesy ESA/Deuros.)

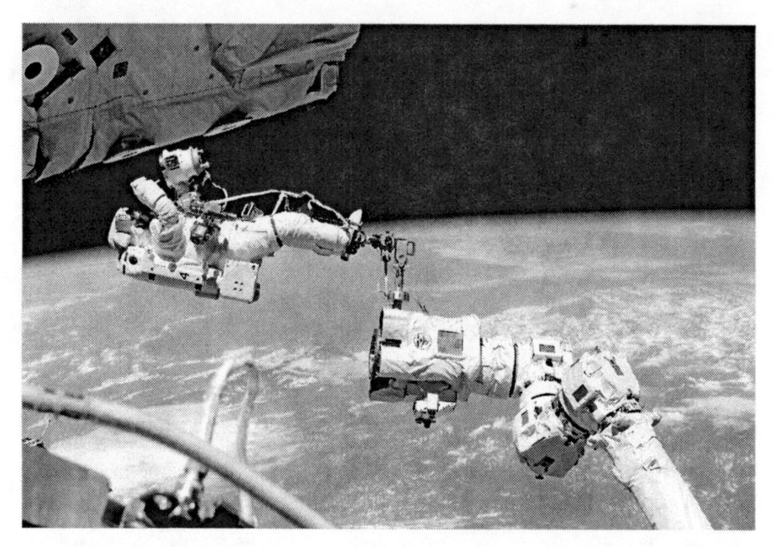

PLATE 12 An astronaut in EVA mode works outside the ISS. (Courtesy ESA.)

PLATE 13 A futuristic view of a Ramjet Fusion Propulsion system. Da full-scale exploration of the Solar System will require the development of advanced propulsion systems. (Courtesy NASA.)

PLATE 14 Astronaut Gene Cernan rides the 'Lunar Buggy' in search of geologically interesting sites during the Apollo 17 mission. Mobility is a key issue in any exploration scenario. (Courtesy NASA.)

PLATE 15 A Mars exploration scenario with a Mars astronaut/explorer scaling a martian cliff in the search for interesting exobiological sites. (Courtesy NASA.)

PLATE 16 The Mars Labs from the ESA Aurora study. (Courtesy ESA.)

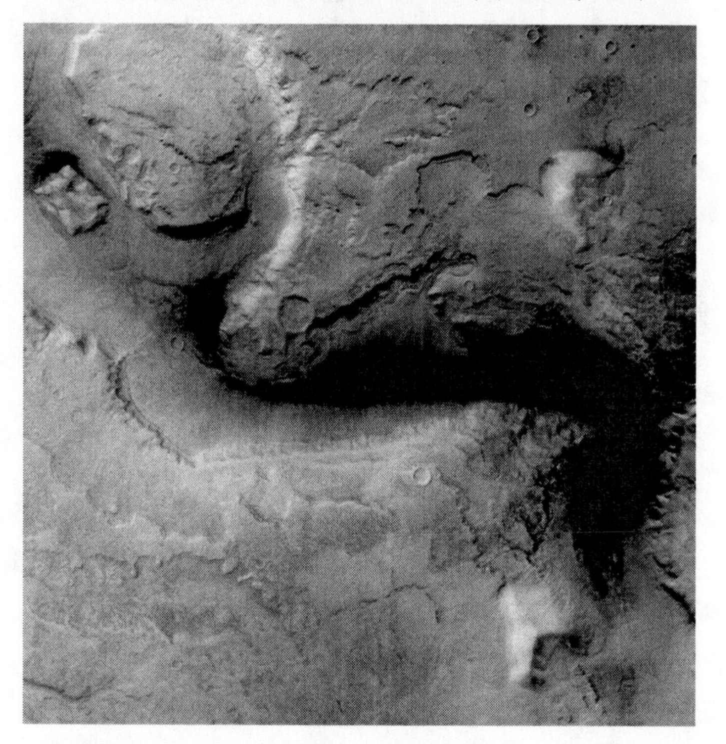

PLATE 17 Reuill Vallis imaged from ESA's Mars Express showing clear evidence of water-sculpted channels, gullies and canyons. (Courtesy ESA.)

PLATE 18 The northern hemisphere of Mars in winter as viewed by the Odyssey spacecraft neutron detectors. The blue regions indicate water ice. In the summer the CO_2 ice melts to reveal the water ice beneath (see below, and also Figure 11.5). (Courtesy NASA.)

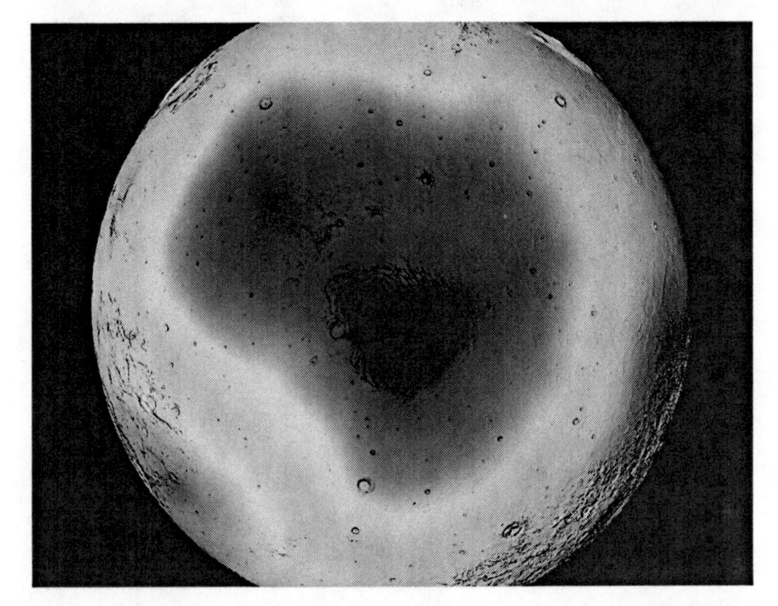

PLATE 19 The northern hemisphere of Mars in summer (also see above). (Courtesy NASA.)

PLATE 20 Artist's impression of EXOMARS, an ESA misson in 2009/11 to search directly for the signatures of life below the Martian surface. (Courtesy ESA.)

PLATE 21 The Spirit/Opportunity rovers. The large vertical mast carries stereoscopic cameras for panoramic imaging and navigation. The forward arm carries instruments for soil analysis such as alpha particle X-ray and Mössbauer spectroscopy and microscopic imaging, to determine the detailed composition of the Martian soil. (Courtesy NASA.)

PLATE 22 Humans and robots on the Martian surface in the search for life. (Courtesy NASA.)

PLATE 23 The Huygens probe, parachute deployed, descends to Titan's surface. (Courtesy ESA.)

PLATE 24 The Rosetta lander analysing the surface of the comet 67P/Churyumov-Gerasimenko in 2014. (Courtesy ESA.)

PLATE 25 Base on Callisto. (Courtesy NASA.)

Part IV
The cosmic biological imperative

... No matter how much we learn about the varied life forms of Earth and the physical nature of the Universe of which we are a part, the question of our biological uniqueness in the Universe is central to the quest for who we are and what our role in nature may be, questions as much a part of religion and philosophy as of science. As Harvard Professor Karl Guthke has claimed in his book The Last Frontier, *the question of extraterrestrial life is one of the most important myths of the modern age.*

Steven J. Dick (1998)

10 The key technologies for human planetary exploration

If even one case of life is shown to exist beyond the Earth, it will transform the way humanity perceives the cosmos and its place in it, and will provide strong circumstantial evidence that life must be a cosmic phenomenon. If no life beyond Earth is found in the Solar System, this too would have dramatic consequences for humanity's view of itself and the cosmos. There arises, therefore, an imperative for us to solve this issue one way or another now that the technology to do that is coming on-line in the first part of the twenty-first century.

When we consider the necessary steps that any comprehensive human exploration programme of the Solar System must go through, we are limited by the constraints of emerging technological capabilities, the inescapable laws of celestial mechanics and the limitations imposed by human economies, social capabilities and goals. Space agencies have long considered that the process will be a 4-stage one. These stages are:

- Orbital, fly-by or crash landing surveying missions
- One-way remote exploration missions to make the first detailed assessment of the planetary surfaces
- Two-way robotic sample return missions, including so-called '*in-situ* resource analysis'
- Two-way human exploration missions to planetary surfaces, first for limited-duration stays, and ultimately with permanent colonies

In order to carry out these mission profiles a number of key technological capabilities must be in place. In the context of human exploration of the Solar System in the search for life, we can identify the four key technologies as:

- Propulsion technology
- Life support and protection technology
- *In-situ* resource technology
- Exploration technology

Put in terms which are perhaps more familiar, propulsion technology is about getting there and back in the quickest and most efficient manner; life support and protection technology is about survival; *in-situ* resource technology is about using what you find there to survive and thrive; and exploration technology is about the exercise of curiosity and intelligence in the acquisition of knowledge and understanding.

PROPULSION TECHNOLOGY

A bold expansion out into the Solar System by sample return robotic missions and by human missions will require equally bold investments in the development of new propulsion technologies. At present, the technologies available operationally to us are chemical propulsion with liquid or solid propellants and solar electric propulsion.

In the case of chemical propulsion, a fuel is burned with an oxidising agent and the resulting gases are exhausted to provide thrust. The classical liquid motor employs liquid hydrogen as the fuel and liquid oxygen as the oxidiser. This is the system employed by the main engines of the Space Shuttle. In the case of solid motors, the fuel and the oxidiser are mixed together in solid form and combustion takes place inside the rocket body, with the combustion gases that result ejected from a nozzle at the back. In the case of electric propulsion, a gas is ionised in a chamber within the motor and then strong electric fields are used to accelerate the ionised atoms out through an aperture in the electrode, as shown in Figure 10.1.

Present-day chemical propulsion can just about support human missions to Mars of the Robert Zubrin type (to be described later). Chemical propulsion can also support non-return missions to Jupiter,

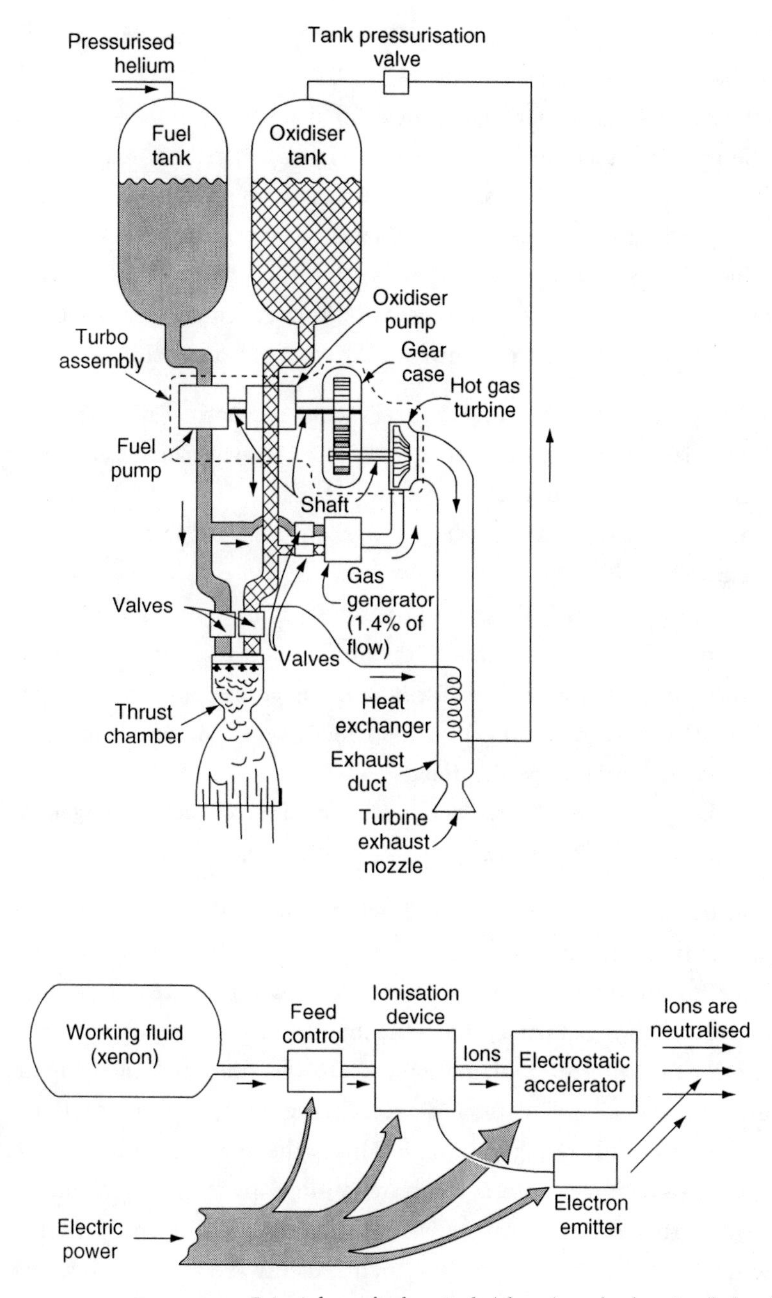

FIGURE 10.1 Principles of chemical (above) and electric (below) propulsion (from *Rocket Propulsion Elements* by G.P. Sutton and O. Biblarz, Chichester: Wiley Europe, 2001).

Saturn and the outer planets with relatively small spacecraft, like the Voyager and Pioneer spacecraft of the 1970s and 1980s. Robotic sample return missions to Mars and Venus can also be just supported. But at that point the technology is very near its performance limits. A recent study by ESA for a Mars sample return mission in 2009 requires two Ariane 5 launches to carry out a mission bringing back a few tens of kilogrammes of soil sample or rocks. To understand this more deeply requires knowledge of the famous 'Rocket Equation', which is explained in Appendix 6. Summarising what is explained there:

- Every target planet or comet or asteroid has a quantity called delta *V* associated with it (it is the increment of velocity needed to get there from the Earth's surface).
- Each rocket technology has a quantity called specific impulse I_{sp} associated with it.
- The ratio of delta *V* of a target planet to I_{sp} of the chosen propulsion method determines the ratio of the final landing mass at the planet to the launcher mass from the Earth – the higher the delta *V* requirement (usually the further away the planet), the lower the mass at landing and conversely for the specific impulse I_{sp}.
- A third quantity – thrust – determines the time to reach the target planet: the higher the thrust, the shorter it takes and vice versa.

The exponential nature of the Rocket Equation tells us that with conventional rocket technologies, missions to Mars are at about the limit for human exploration. To do a 4-crew human return mission within existing Saturn V/Shuttle technologies even requires the production of the return fuel (methane) on the Martian surface, using the *in-situ* atmospheric carbon dioxide along with a store of liquid hydrogen already brought to Mars. This is the Robert Zubrin scheme.

How can we contemplate human missions in the search for life beyond this type of Mars mission? Clearly, *in-situ* production of return fuel will be a key requirement; also it seems clear that new propulsion technologies will be decisive. But what will these new propulsion technologies be?

The candidates are:

- Nuclear (fission) electric propulsion
- Nuclear (fission) thermal propulsion
- Nuclear fusion propulsion

Viewed from the perspective of the year 2003, the first two of these have a very good chance of being developed and used within the next 20 to 30 years, whereas the third has an entirely unforeseeable future.

Without doubt, human exploration of the Solar System beyond Mars will have to be based on nuclear technology. The rocket technology of the present, based on chemical rockets and subject to the exponential tyranny of the Rocket Equation, dictates this, as is shown in Appendix 6. There are two ways nuclear fission power can be used. The first is by the use of radioisotope generators that produce electricity from the heat released by the radioactive decay of plutonium-238. Usually this heat is applied to a thermocouple that produces electricity at about 6 % efficiency. But if the thermocouple is replaced by a Stirling engine this can be increased to 23 %. Here the nuclear reactor is used to heat a fluid. The hot fluid is then fed into the Stirling engine, where the heat is used to drive a piston that in turn drives an electric generator. The electrical power is then used to drive an ion thruster in which a heavy gas, such as xenon, is ionised and the electric power accelerates the ions out through a nozzle to generate thrust. The Jupiter Icy Moons Orbiter, described in Chapter 11, is planned to be powered by such a motor.

The next technology is the nuclear thermal technology. Here the heat generated by a nuclear reactor is used to heat up a propellant gas, such as hydrogen (delivered from a liquid hydrogen reservoir), to high temperatures so that when it is expelled from a nozzle at the back of the motor it produces enormous thrust. These nuclear thermal motors are the most powerful rocket motors devised to date. They will be fully capable of delivering the propulsion necessary to carry out both robotic and human exploration missions throughout our Solar System. The development of this type of nuclear rocket propulsion

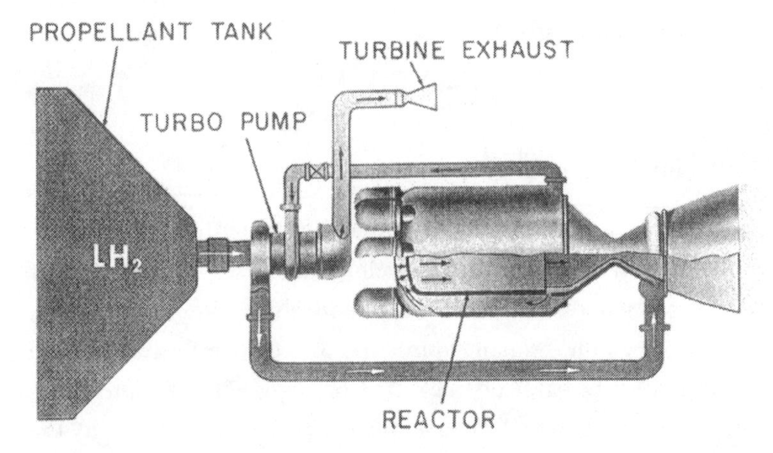

FIGURE 10.2 A nuclear electric propulsion system. (Courtesy NASA.)

technology is fundamental to establishing our role in the intelligent exploration and search for life in the twenty-first century.

Given the political, technological and cultural realities and constraints, the timetable for the development and implementation of the key technologies indicated above could be:

- Nuclear electric 20 years
- Nuclear thermal 50 years
- Nuclear fusion 100 years

In February 2003, NASA announced the start of the $3 billion Project Prometheus that will develop a new generation of nuclear propulsion systems for missions to the great and outer planets. The project will be undertaken jointly with the US Department of Energy, which will be responsible for the nuclear fission part of the work. The initial thrust of the programme will be towards nuclear electric propulsion. Nuclear thermal technology, although tested by the USA in the 1960s, attracts environmental criticisms and has a longer term prospect.

Finally, in terms of the perspectives of what might be possible within this century, and bearing in mind that controlled nuclear fusion power on the ground could be within our grasp in the coming decades, it seems plausible to extend that capability to space. The controlled fusion of hydrogen nuclei, as in the hydrogen bomb or the interior of the Sun, will provide rocket motors of enormous thrust.

A ramjet fusion propulsion system is shown in Plate 13.

LIFE SUPPORT AND PROTECTION TECHNOLOGY

If human beings are to have a real chance of surviving in the longer term in space conditions outside the benign conditions at the Earth's surface, guaranteed by our atmosphere, then sophisticated technological means well beyond our present capabilities will need to be deployed. From a purely technological point of view, these means consist of three elements:

- Open life support systems
- Closed life support systems
- Life protection systems

Up to the present time, all space life support systems are of the open variety. That means that the oxygen, water, food and other consumables are brought into space, consumed by the crew and the waste products returned to Earth or vented to space. There have been some exceptions to this, such as the recycling of grey water (e.g. waste shower water) developed by Russian engineers for the MIR space station; but by and large the burden of supplying the bulk of consumables for life support has had to be borne by logistics re-supply from Earth.

With the establishment of long-duration stations in Earth orbit, beginning with Skylab in the seventies, then the construction of the Salyut and MIR stations, and finally the International Space Station, experience has gradually grown in equipping 'homes in space'. For these space stations, the cargo must include not only food and personal belongings, but also the air to breathe and the water to drink or to be used for hygiene and cleaning purposes.

In order to provide such a home in space, the human requirements need to be specified. These requirements are:

1. the design of the spacecraft to protect the crew from the hostile environment of outer space,
2. the quantity and quality of consumables needed, including oxygen, food and beverages, potable water, water for personal hygiene, hygienic articles and clothing,
3. the management of waste produced with regard to solid matter, including trash, liquids and gas contaminants production, and
4. the management and quality control of this closed environment.

It turns out that water is the major driver in meeting human needs. Depending on the mode of water usage for hygiene purposes, either getting by with soap-impregnated towels, wet and dry towels, dry shampoo and chewing gum for cleaning purposes, or assuming one shower per day per person, the total amount of water needed has been assessed as 4.6 kg per person per day (with dry to humid cleaning only), or 25.8 kg per person per day (including a daily shower). Hence, water may contribute to roughly 80 % of the human needs and waste (Figure 10.3). The average daily oxygen consumption is roughly 1 kg per person, under the assumption that the astronauts undertake intense physical training in order to keep their body functions in a healthy state. The total needs and wastes therefore amount to 30 kg per person per day.

So far, for space stations such as MIR and the ISS, consumables have been/are carried from Earth and the wastes either stored or loaded in trash spacecraft that are burned in a destructive re-entry.

Transportation, bringing the consumables required on board and discarding the waste, is the first step in building the home in space. As in every confined habitat where human beings live and work for extended periods of time, a life support system is needed to provide and maintain a physiologically acceptable environment. This includes the management of the resources required for humans and other biological species, as well as of the human and other wastes. Experience gained from the use of life support systems on submarines has also been useful for space application.

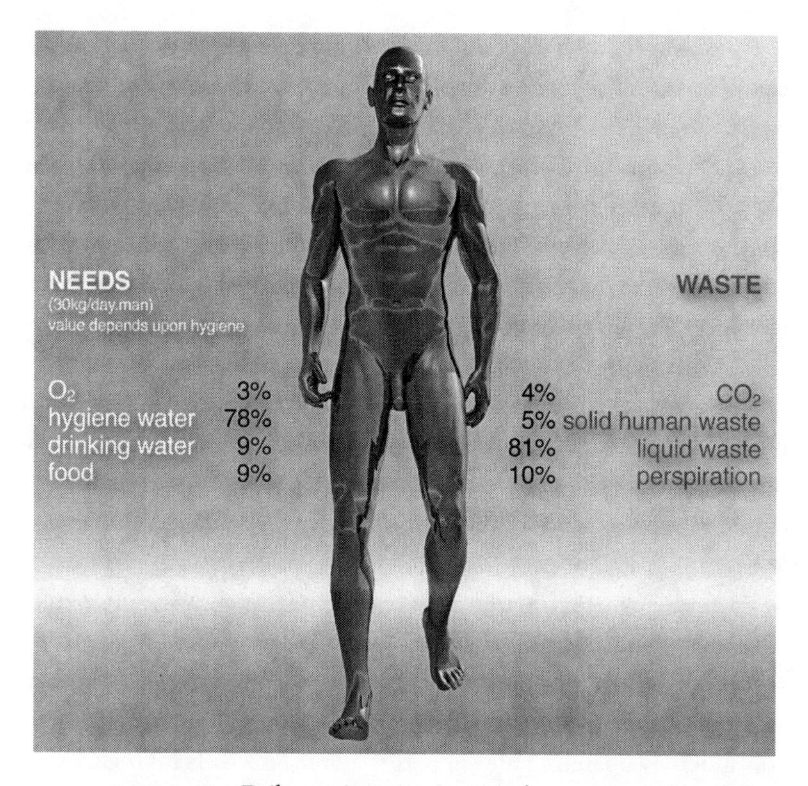

FIGURE 10.3 Daily requirements in space for one person (sketch from Space Channel).

The life support technologies used in the two major manned space stations in Earth orbit, the Russian space station MIR and the International Space Station, integrate physical and chemical technologies. In both systems, food and nitrogen are supplied from Earth. Oxygen is also supplied from Earth, either as compressed gas or released from water by electrolysis. Water, being the largest part of the human requirements (Figure 10.3), is to a large extent recycled. Different techniques have been developed to recycle waste water, such as osmosis or reverse osmosis processes, filtering through specific membranes, distillation, and/or ultrafiltration or catalytic processes. Some of these technologies have already been installed in the stations in space, thereby reducing substantially the upload mass and related operational costs.

However, the safety and maintenance concepts of the ISS, for example, strongly rely on regular re-supply flights, even in the presence of recycling systems. Such ouposts in space as the ISS benefit from the short distance from their orbit to the Earth surface, allowing fast transportation and replenishment of stocks, and even rapid evacuation of the crew in the case of severe problems. Today, the life support system on board the ISS allows for a permanent crew of seven members living and working in space for several months.

Considering a lunar base, the human needs and life support requirements are similar to those of a space station in low Earth orbit. The Moon can even be considered as a natural space station orbiting the Earth at an average distance of 384 400 km, which is of course much further from the Earth than the ISS, with an orbit ranging between 325 and 400 km. For the transfer phase from the Earth to the Moon and back to Earth, lasting only 3 to 5 days each, all consumables necessary to keep the crew alive would be available on board the spacecraft. When arriving at the lunar base, consisting of a permanently equipped and inhabited complex, the crew would rely on a continuously operating life support system. This calls for highly reliable and stable systems.

The requirements of a crew of four, living and working on a lunar base for 180 days, are significant. The consumables needed amount to 22 tons, composed mainly of water (18.88 tons assuming a daily shower), food (about 2 tons) and oxygen (0.73 tons). Even assuming that the regular transport capacity would replace a quite high proportion of the necessary consumables, it would be advantageous to close the life support loops to the greatest extent possible with regard to the atmosphere, and, even more importantly, with regard to water. The proximity of the lunar base to the Earth will allow a step-by-step setting-up of more and more complex life support systems. In a first step, efforts will focus on the recycling of the atmosphere and of water that comprise about 90 % of the mass of consumables for a 180-day mission.

These physical-chemical processes can be complemented by so-called bioregenerative methods. Since the global cycles of oxygen, carbon dioxide and water are driven by the activities of life itself, bioregenerative processes can recycle resources, thereby mimicking the global cycles that sustain our own biosphere. In preparatory studies, algae have demonstrated their capacity to manage the atmosphere in confined habitats by consuming carbon dioxide and producing oxygen. In the BIOS 1 to 3 Ground Experimental Complex, the research group led by Josef Gitelson at the Institute of Biophysics of the Russian Academy of Sciences in Krasnoyarsk has succeeded in sustaining an 80 % to 90 % closure of a human habitat over half a year. In these complexes, the alga *Chlorella* served as an element for gas regeneration, water treatment and even for food production.

Algae are known to play a significant role in the photosynthetic restoration of organic substances of our own biosphere: they produce 5.5×10^{10} tons of biomass per year, which amounts to 33.5 % of all organic substances annually synthesised in the biosphere (1.64×10^{11} tons). Due to their consummate autotrophism (they need only sunlight and carbon dioxide for their growth), their simple and asexual method of reproduction and their broad ecological adaptability, algae are the best candidates for bioregenerative life support systems in habitats beyond the Earth.

In addition, biological water treatment systems, for example, the treatment of crew waste by bacteria and perhaps fungi, could be used for water regeneration/production, followed by a physicochemical treatment depending on the final usage of water for hygiene or drinking purposes. However, even assuming a perfectly functioning bioregenerative life support system, physicochemical processes should always be available as back-up technology. In a later more advanced step, inflatable greenhouses could be used to produce vegetables as supplementary food. In such greenhouses on the surface of the Moon, the natural resources of the Moon will be used to the greatest extent possible, such as the natural sunlight, the lunar soil and/or even the water, if frozen water is found in the shaded areas of the polar craters. Species will be selected that have been commonly used as food plants providing the main nutritional needs of humans.

In this context, wheat, soy beans, lettuce, and potatoes, but also carrots, beet, radish and cucumber have been proposed. Furthermore, there is the psychological value of growing plant cultures in an otherwise hostile environment which should not be neglected.

At the end, the life support systems developed in the different phases will operate together to close the life support loop to the greatest extent possible. Algae are more manageable than higher plants for the control of the oxygen/carbon dioxide balance, and microbial and fungal compartments are necessary for the recycling of the inedible biomass produced by plants, such as roots, straw, etc. Hence, the Moon is a perfect location to test and evaluate complex self-sustainable life support systems without raising major safety issues, since relatively rapid emergency transport will exist from and to the Earth.

The demands on life support systems increase drastically, however, whenever humans take part in exploratory types of missions involving interplanetary and planetary environments. The human needs for a mission to Mars, assuming a 1000-day voyage of a crew of 6 with 4 members staying on the surface of Mars for about 500 days (Table 5), amount to about 108 tons for the transfer periods between Earth and Mars and 65 tons for the crew working on Mars. For comparison, the payload of the first European lander Beagle 2 (named after the vessel that was used by Charles Darwin on his exploratory voyage, see also Chapter 1) is limited to about 10 kg, less than one ten-thousandth of the above needs. No currently available vehicles are capable of transporting these enormous loads to Mars. An additional constraint faced by the crew is that after having left low Earth orbit no space port can be called at to replenish resources. Therefore, to satisfy the human needs on missions to Mars, the transfer vehicle, as well as the outpost station on Mars, must be equipped with life support systems capable of recycling as much of the waste products as possible.

A life support system capable of recycling the atmosphere and water would reduce the mass of consumables by 90 %, i.e. in this case, from 108 tons to 11 tons. Complementary to physicochemical systems of regeneration, the most promising biological systems are algal and

TABLE 5 Life support mass figures for 1000-day Mars missions.

Consumables and products	Transfer phases in total (tons)	525-day stay (outpost) phase (tons)
Oxygen	3.65	2.15
Hygiene water	82.40	48.00
Potable water	10.00	6.00
Food	9.60	5.60
Carbon dioxide	4.45	2.60
Water vapour	10.60	6.20
Solid waste	7.50	4.40
Liquid waste	85.60	50.20

microbial reactors. This is also because of their relatively small size, their dynamic response time, and the possibility to control them and – in case of failure – to restart the system relatively fast, i.e. within a few days.

If space is available, higher plants could be introduced as sources of fresh food complementing the diet by up to 30 %. However, the area required for a plant culture is large; an area of about 15 m^2 is estimated as the minimum requirement for a monoculture sufficient to feed one person. Hydroponic cultures are suitable modes for the cultivation of higher plants in low gravity. With the addition of plants, the oxygen and water might be recycled by nearly 100 %, and 30 % of the food could be provided. This would lead to a further reduction of the mass of consumables for the transfer and the Mars orbiting vehicle to about 6 tons. However, in addition to the human needs, the equipment and instrumentation needed for the bioregenerative systems will add to the cargo. These numbers are already in the range of technical feasibility, since the current Space Shuttle fleet has a maximum of 30 tons payload capability.

Upon arrival at Mars, the lander and habitat module will bring four members of the crew to the surface to live and work there for more than a terrestrial year. Already, some of the cargo will have been launched in advance by robotic missions. The nearly Earth-like day/night cycles (a Martian day, called sol, lasts for 24h 37′ 22.7″) will favour plant growth under direct sunlight. The structures required for setting up the cultivation will have already been installed, such as inflatable structures for greenhouses and a biological life support system that can be quickly started. The crew needs just to bring the seeds and stock cultures. Unicellular organisms, such as algae, bacteria and/or fungi are the most probable candidates for starting the bioregenerative life support system, which can then be complemented further by higher plants to produce fresh food. In any case, the natural resources of the red planet will be used to the highest extent possible. These are:

(i) sunlight for power production and as an energy source for photosynthesising cultures,
(ii) atmospheric compounds, especially carbon dioxide, constituting 95 % of the Martian atmosphere,
(iii) water which forms large ice deposits at the poles, but exists also as frozen ground water in the permafrost regions below the surface, and
(iv) Martian regolith as soil for plant cultures, as source of volatiles to be extracted, and finally for radiation shielding purposes.

In addition, nuclear power plant(s) brought to Mars prior to the arrival of the crew will substantially reduce the energy problem. At the end of the Martian excursion, the plants installed at the surface of Mars will produce a valuable portion of consumables required for the return mission from Mars to Earth.

EXPLORATION TECHNOLOGY

The act of exploration itself on planetary surfaces such as Mars, Europa and comets, the means of landing, achieving mobility and the process of retrieving samples, performing geological surveys and the analysis of samples after drilling will all require technologies expressly developed for the purpose. The symbiotic relationship between human explorers and smart robots will be crucial here.

The key technologies will be:

- Surface rovers and human mobility vehicles
- Surface robots
- Search for life sensors/analysis technologies
- Drilling technology
- Planetary surface laboratories
- Sample preparation and return technologies

Surface rovers and human mobility vehicles

Automated surface rovers for investigating lunar and planetary surfaces started doing this with the Russian Lunakhod lunar rover of the 1970s. On Mars, the first real rover was with Pathfinder of 1998, but this was still a small unit with a slow mobility confined to a few metres around the landing site. Larger and more mobile units came with the two Mars Exploration Rovers (MER) landing in early 2004. This type of rover consists of:

- **a body:** a structure that protects the rover's 'vital organs'
- **brains:** computers to process information
- **temperature controls:** internal heaters, a layer of insulation
- **a 'neck and head':** a mast for the cameras to give the rover a human-scale view
- **eyes and other 'senses':** cameras and instruments that give the rover information about its environment
- **arm:** a way to extend its reach
- **wheels and 'legs':** parts for mobility
- **energy:** batteries and solar panels
- **communications:** antennas for 'speaking' and 'listening'

In some senses, the rover's parts are similar to what any living creature would need to keep it 'alive' and able to explore.

This type of rover extends considerably the mobility and exploration capabilities but to a lesser degree the scientific search-for-life experimental analysis capabilities as the science instruments are still fairly rudimentary. In this sense, this type of rover is probably a precursor of the type of roving, mobile and semi-intelligent robot that humans will need to support their planetary surface exploration on surfaces such as Mars, where the intelligent robot can access difficult and possibly dangerous but potentially interesting terrains for exobiology.

More sophisticated, automated search-for-life analyses will be carried out by rovers which combine medium to high mobility with sophisticated drilling, sample retrieval and analysis capabilities. These more sophisticated analysis techniques were to be inaugurated by the tiny Beagle 2 lander of early 2004 that, without any rover capability, nonetheless would have carried out on the landing spot mass spectrometry analysis for organics, coring and grinding, and other analyses. But it is the landers of the latter part of the decade, like ESA's presently planned EXOMARS mission, that will place on the Martian surface rovers that are sufficiently mobile and analytically capable that they are likely to go a long way to answering the question of whether life is or ever has been present on or just beneath the Martian surface.

From the point of view of mobility and the extension of developed technologies for human surface exploration we have to go back to the early 1970s to find the predecessor of what will be the future human mobility. On the Apollo 17 mission in 1972, Gene Cernan and Harrison Schmitt, a trained geologist, showed what they could do in terms of wide-ranging geological research when supported and made highly mobile with the help of the 4-wheeled Lunar Rover Vehicle (LRV).

During the development of the Mars Reference Mission of NASA for human Mars missions in 1997, a degree of mobility on several scales was considered necessary. Crew members outside the base would be in pressure suits and could operate within walking distance served by rovers, carts and wagons. Using unpressurised wheeled vehicles, ranges between 1 and 10 kilometres could be achieved. Beyond that, pressurised rovers would be used, allowing a shirtsleeve environment for the crew. For long-range exploration up to 500 kilometres, 10-day sorties were considered, with half the excursion time used for travel and 16 person-hours per day available for EVA. The rovers would use a nominal crew of two but have an emergency capability of carrying four.

Surface robots

As we have seen, surface robots will be used to support human explorers in the process of exploration of planetary surfaces. Therefore, the emphasis for these machines will be on the capabilities of mobility and access to difficult terrain in support of human search-for-life exploration. Starting with the Lunakhod robot of the Soviet Union of the 1970s, an evolution into two types of robot can be foreseen.

The Viking/Beagle 2 landers can be characterised as automated science packages and the Pathfinder/MER rovers as semi-intelligent explorers. One of the major functions of the latter will be the remote (from humans) mapping of difficult or inaccessible terrain. Recent developments in robotics have started to make significant progress in this difficult problem for robots. Up to now, robots have been developed either to accurately map their environments or to follow existing maps. Doing both is presently difficult for robots. Even doing both separately has its problems. Following a map requires accurate tracking of where the robot has been, and calculation of the distance and orientation to the starting point, i.e its present position relative to the map. The slipping of wheels and the execution of inaccurate turns all add to the cumulative error. Making a map is equally difficult and requires the ability to recognise features already encountered. Yet making a map is the primary act of exploration. Recently a team at NASA's Ames Research Center found that with a new type of algorithm great progress can be made on both problems by solving them simultaneously. The technique is called Simultaneous Location and Mapping, or SLAM. Based on these advances, it is clear that semi-intelligent, wheeled robots will play a significant and probably the major role in the detailed on-the-surface mapping of Mars and other planetary bodies.

Search-for-life sensors/analysis technologies

The main scientific objective of search-for-life strategies is to search for chemical and biological indicators of past and/or present life. This includes drilling and sampling the subsurface in the case of Martian soil. It would also include the use of several dedicated subsystems to perform:

- elemental analyses,
- mineralogical determination, and
- isotopic and molecular analyses of inorganics and organics in connection with structures and life processes.

In the case of Mars, and possibly on other bodies, it is essential to have access not only to surface but also to subsurface samples to a depth where:

- the possible effect of UV radiation on the chemical indicators of life is negligible; and
- the concentration of oxidising agents such as H_2O_2, a potential source of degradation of organics, including bio-organics, is negligible.

The sample preparation equipment therefore would include:

- a grinding device to take off the crust of rock boulders and smooth the surface of boulders and drilled cores
- a core drilling device to probe the interior of hard rocks and regolith
- a fine grinding/polishing device to smooth the ground surface for optical investigation and microanalysis
- a magnetic device to separate minerals/grains with a variety of magnetic susceptibilities (ferromagnetic, paramagnetic, diamagnetic)
- a sampling device to allow samples to be taken from flat ground or polished surfaces (small core drill, chipping, etc.)
- a sample transfer system (manipulator) to transfer prepared samples and selected grains to the chosen analytical instrument

The pursuit of this type of strategy probably represents the optimal approach to the detection of traces of extinct or extant life in these planetary environments. The detailed sample analysis techniques and technology are described in Appendix 5.

Drilling technology

Extensive terrestrial experience exists with industrial-type drilling technology, such as exists in the oil industry. Exploratory drilling on planetary surfaces will be something between this and dental surgery. In fact the drill/corer on the ill-fated Beagle 2 lander on Mars (see Chapter 11) has been developed by a dental surgery team.

In ESA's Rosetta mission to Comet 67 P/Churyumov-Gerasimenko a sophisticated drill will be used to drill into the comet's surface. A similar drill will be used on the ESA EXOMARS mission on Mars to drill into the Martian surface up to 1.5 to 2 metres to reach below the supposed oxidising layer. These types of drill will have drill bits of about 1 cm diametre and perform drilling by cutting. Samples about 0.5 to 1 cm in length would be retrieved and samples could be taken at multiple levels in the borehole, the number limited by drilling speed, the energy available for the drilling programme and the speed at which samples can be processed. In the case of purely robotic missions, this last point may be limited by the quantity of data produced by each sample and the limitations on transferring the data back to Earth. The drill power unit will need to provide power for thrust, torque and percussion. In the case where drilling is into solid rock rather than regolith, the drill bit will need to be replaced by a rock corer. A carousel system is probably required to store retrieved samples and act as a system to deliver in a sequenced fashion the sample or fractions of samples to the various analysis instruments, such as those described in Appendix 7.

Sample handling and preparation technology

Sample handling and preparation for analysis is extremely important. Without being carried out properly (leading to ambiguous or false results), it would jeopardise the mission in the case of robotic missions

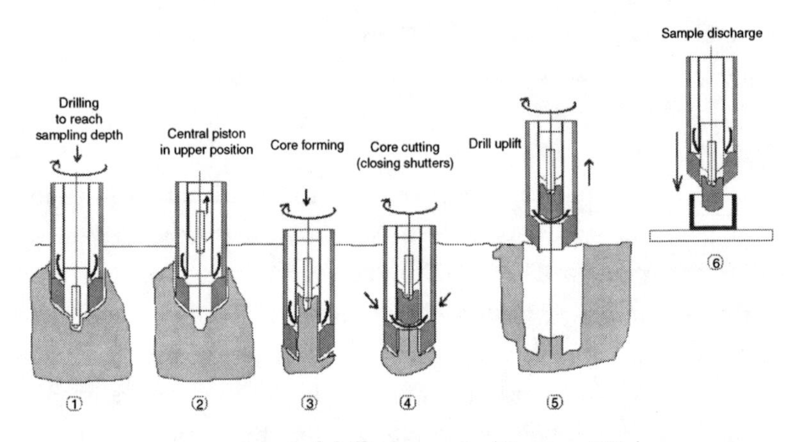

FIGURE 10.4 A typical drilling scenario. (Courtesy ESA.)

or field trips in the case of human exploration on planetary surfaces. The key functions to be carried out are:

- Start with samples as solid rock cores or as regolith gravel
- Reduce material to fine powder less than 10 microns for delivery to pyrolysis and chemistry analyser
- Produce smooth, polished surfaces for optical examination
- Produce material 0.2 mm deep with mica windows for Mössbauer spectroscopy

In conclusion, it can be stated that optimised drilling, sample preparation and analysis are mandatory in search-for-life researches on planetary surfaces if we are to avoid the sort of ambiguous results which were found in the relatively expensive Viking missions of the 1970s. This is particularly true for robotic missions. In the case of human missions, where the upfront sample acquisition, preparation and analysis can be carried out or at least supervised by humans, there is the possibility of detecting on-the-spot anomalies or ambiguities and correcting for them.

Planetary surface laboratories, e.g. a surface laboratory on Mars

Various studies have looked at the optimum configuration of surface habitation and laboratory modules on Mars. We will look at two of them: (i) the NASA Mars Reference Mission modules and (ii) the ESA study laboratory.

In the NASA Mars Reference Mission, a 2.5-metre-diametre cylindrical module on two levels was considered, with one level being a non-sensitive stowage element housing crew support elements and the other level carrying the primary science and research lab. The module is considered to be able to support:

- fieldwork
- tele-robotic exploration
- laboratory and inter-vehicular experiments
- preparation of samples to be returned to Earth

The NASA study estimated that as experience grows, local human exploration will be extended to more regional forays lasting several weeks and using mobile facilities, and may be conducted at intervals of a few months. Between exploration forays, analyses in the laboratory will continue. The surface laboratory would be co-located with a similarly structured habitation module or modules on the Martian surface.

As far as European thinking in this area is concerned, we can cite a recent study for ESA (European Mars Missions Architecture Study) that identified two elements that could be provided by Europe, namely:

- A Mobile Pressurised Laboratory (MPL)
- A fixed laboratory or Biology/Greenhouse Module (BGM)

The MPL would have a range of 500 km and mission duration of 20 sols (Martian days) before refuelling/recharging at the *in-situ* resources utilisation (ISRU) facility (see Plate 16).

It would accommodate two crew during that period and would be capable of travelling over the ground at 5 km/hr. It would have a surface EVA airlock allowing the crew to exit to the Martian surface to conduct geological and exobiological investigations. To aid this there would be an externally mounted science glovebox and robotic arm. The MPL would be somewhat larger than a large camper van and could also provide a safe haven capability at a Martian base for up to six crew members.

As far as the BGM is concerned this would provide four functions:

- An exobiology and *in-situ* resources research capability
- A biology technology testing capability
- A food production, preparation and storage capability
- A Closed Environmental Life Support System (CELSS)

The greenhouse would be used for growing plants as food but would be used also as one of the elements of the CELSS.

IN-SITU RESOURCE EXPLOITATION TECHNOLOGY

The case made by Robert Zubrin for *in-situ* resource support for human missions to Mars, and in particular the case for the use of local atmospheric carbon dioxide to produce the methane fuel for the return journey to Earth (see next page), seems overwhelming. In general, the list of resources that can be exploited are:

- CO_2 from the Martian atmosphere to produce methane (CH_4) as fuel and water for crew, as well as a source of hydrogen (H_2) and oxygen (O_2)
- Oxygen from the Moon as an additional fuel to hydrogen in nuclear thermal propulsion
- Planetary (for example, lunar/Martian) soil as a growth medium for plants/food
- Planetary atmospheric gases for various purposes
- Planetary minerals
- Planetary micro-organisms (if found) for biotechnology
- Sunlight for photosynthesis and photovoltaic power generation
- Planetary water found at the lunar poles and in subsurface Martian reservoirs to be used for crew consumption and hygiene
- Helium-3 from the Moon as a fuel for fusion reactions

The strongest proposal yet for the use of *in-situ* resources comes from the work of Robert Zubrin. Faced with the tyranny of the Rocket Equation and the fact that 90 % of the total mass of spacecraft on launch from the Earth's surface is propellant, Zubrin proposed that in the case of Mars missions, a significant proportion of propellant for the return to the Earth could be manufactured on Mars. This then removes the need to transport this propellant from the Earth to Mars in the first place and dramatically alters the launch mass requirements. The basic process involves the reaction of hydrogen with carbon dioxide from the Mars atmosphere to form methane and water, according to the reaction:

$$CO_2 + 4\,H_2 \rightarrow CH_4 + 2\,H_2O$$

This is known as the Sabatier process, which is well understood and has been widely used since the nineteenth century. The water so produced can then be further electrolysed to form molecular oxygen and hydrogen according to:

$$2\,H_2O \rightarrow 2\,H_2 + O_2$$

There are four products of these reactions, namely methane, water, hydrogen and oxygen. The methane is then stored as the propellant with the oxygen for the return journey. Some of the oxygen can be also used in the life support system and some of the water can be used to sustain the crew. The hydrogen can be fed back into the Sabatier process. The combined Sabatier/water electrolysis process as it could be employed on Mars is shown in Figure 10.5.

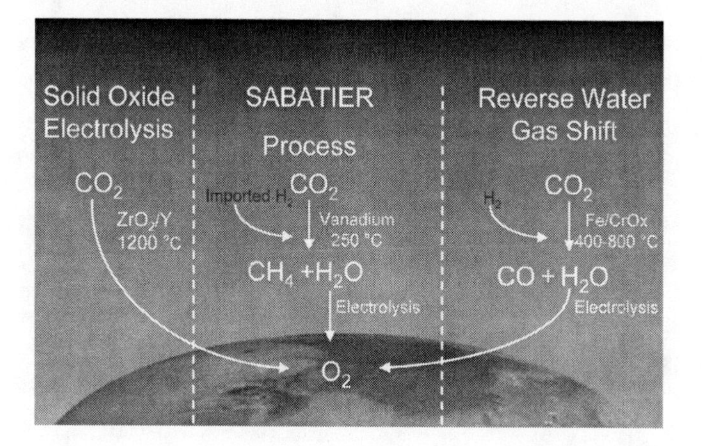

FIGURE 10.5 The Sabatier process (from HUMEX study ESA SP-1264).

Whereas the Moon and Mars are entirely different regarding volatiles on their respective surfaces, both offer different possibilities for *in-situ* resource use. Mars has a thin atmosphere, with 95 % carbon dioxide that can be used to produce propellants and life support consumables as shown above. The Moon, however, has a great store of oxygen at 43 % by weight, but it is bound up in silicates and mineral oxides. Nevertheless, there are chemical techniques that can be used to extract the water and other useful products. The two most favoured are (i) the reduction of ilmenite soil or volcanic glasses by hydrogen at 1000 °C and (ii) the carbothermal reduction using methane at 1600 °C.

The ilmenite/volcanic glasses' reaction is:

$$FeOTiO_2 + H_2 \rightarrow Fe + H_2O + TiO_2$$

And then electrolysis:

$$2\,H_2O \rightarrow 2\,H_2 + O_2$$

Although this process is relatively simple, the yields are rather low, i.e. 3–6 % oxygen by weight. Nevertheless, it was favoured by NASA

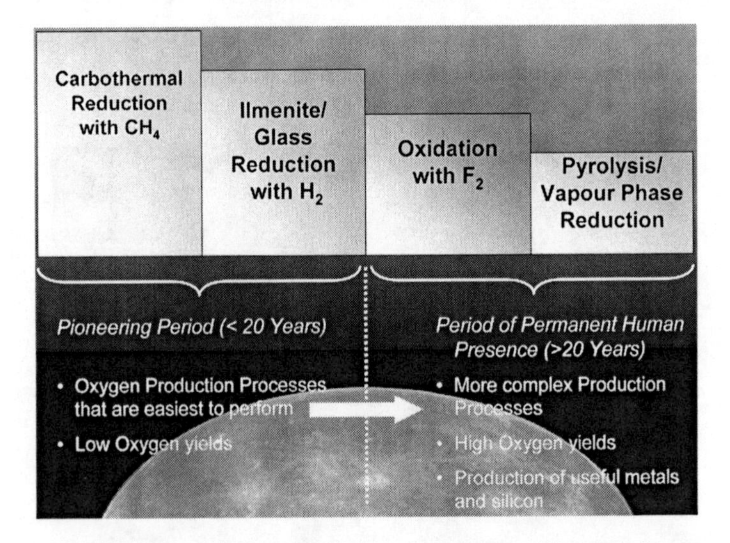

FIGURE 10.6 *In-situ* resource utilisation (from HUMEX study ESA SP-1264).

in its Reference Mission and it has already been proven in an industrial context. The soil used comes from the mare basins or highlands and must be rich in iron oxides. To be efficient, the hydrogen from the second step must be fed back to the first step in a closed loop configuration. This has not yet been demonstrated.

The carbothermal reduction using methane is:

$$Mg_2\,SiO_4 + 2\,CH_4 \rightarrow 2\,MgO + Si + 4\,H_2 + 2\,CO$$

Followed by:

$$2\,CO + 6\,H_2 \rightarrow 2\,CH_4 + 2\,H_2O$$

and then electrolysis:

$$2\,H_2O \rightarrow 2\,H_2 + O_2$$

This process is relatively efficient with an oxygen yield of about 20 % by weight. Again, in order to be efficient, the methane from the second step must be fed back to the first step and the hydrogen from the third step must be fed back to the second step in a closed loop configuration. This has yet to be demonstrated.

These processes will of course require infrastructure on the lunar and Martian surfaces. Estimates of the sizes of these infrastructures have been made in the course of the HUMEX study. In the case of the methane production on Mars, infrastructure including cryogenic storage, equipment for compression of atmospheric gases, processing devices such as chemical reactors, heating devices and electrolysis equipment will all be needed. The total plant would be 9 tons and would produce 5.5 tons of methane and 20 tons of oxygen as propellants per year, along with 5 tons of oxygen and 22.5 tons of water for life support. In the case of the lunar production of oxygen, in addition to the infrastructure elements similar to those cited above for Mars, infrastructure elements for handling and processing the lunar soils would also be needed. For a total plant mass of 50 tons, a yearly production rate of 100 tons of oxygen was estimated.

As has been indicated in the section on life support systems, Moon and potentially Mars soils have been proposed in combination with water as growth media for edible plants such as wheat, soy beans, lettuce, potatoes and other vegetables. The water could be derived from polar ice caps, underground aquifers or from some of the chemical processes for producing methane on Mars or oxygen on the Moon, as described above. Lunar and Martian soil has also been proposed as material to be used in constructing shielding structures against the cosmic radiation reaching the surface of these bodies.

Planetary atmospheric gases are also an *in-situ* resource. For instance, on Mars we have seen a carbon dioxide concentration by volume of 95 %, but there is also 2.7 % nitrogen and 1.6 % argon, as well as oxygen and water in trace amounts.

Planetary minerals are also an obvious *in-situ* resource. We have seen above how the two reactions to be used to extract lunar oxygen produce iron or silicon. Other reactions to extract other important minerals are clearly possible.

If micro-organisms are present in the Martian soil, based on the terrestrial DNA/RNA scheme, they may prove useful, as micro-organisms on Earth do, in many biochemical or biotechnology processes involved in the running of the Martian base. In this connection, it can be noted that the Polymerase Chain Reaction (PCR) universally used today in the processing of DNA was originally developed from the DNA of an extremophile organism.

The use of water derived from permafrost also seems obvious but would clearly depend for its usefulness on the ease of access and extraction. Polar regions may not be accessible from landing sites or bases away from the poles and underground aquifers may be deep and difficult to drill down to.

Lastly, a proposal that is not a new one but still has a high potential, particularly in the longer term, is the mining of the helium isotope ^{3}He on the Moon for use in fusion reactors. Due to the direct impact of the solar wind on the Moon, considerable quantities of ^{3}He are embedded in the lunar soil. In 1986, L. J. Wittenberg and others estimated that this could amount to 1 million metric tonnes and proposed this as a source of fuel for deuterium–^{3}He and ^{3}He–^{3}He reactions for future electricity generating fusion reactor power stations on Earth. The former reaction releases only 1.5 % of its energy in the form of neutrons and the latter none of its energy. This is to be compared with over 50 % for the presently planned D–D and T–T fusion reactors. The problem is that the neutrons end up in the reactor casing/shielding, making it radioactive, thus leaving a heritage of radioactive debris from the supposedly 'clean' fusion reactions. So reactors based on the ^{3}He–^{3}He reactors would be truly clean. However, the ignition temperature and confinement requirements are higher than for the reactors presently planned and so the perspective is clearly longer term, although the extraction and transportation technologies are available today.

In conclusion, a wide range of *in-situ* resource utilisation options present themselves to aid us in the exploration of the planets, some more near term and some more longer term awaiting development and validation in a space context.

11 Exploration in space

This chapter deals with how the exploration will progress in space: starting with the inner planets and the robotic and human exploration of Mars and the polar regions of the Moon in the search for life, progressing to the moons of Jupiter and Saturn, the comets and asteroids, and finally the outer planets of the Solar System, Uranus, Neptune and Pluto. The chapter covers in some detail the initially robot dominated but ultimately robot supported human exploration of the Solar System.

THE INNER PLANETS

> *The inner planets provide a unique opportunity to study the processes that lead to habitable worlds. Venus, Mercury, Mars and the Moon each hold clues to different aspects of the origins of the planets and habitable environments in the inner solar system. The Moon and Mercury are 'Rosetta stones' in that they preserve records of past events that are largely erased on Earth and Venus.*
>
> US National Academy of Sciences (2003)

MERCURY

In 2011 or 2012, the ESA/Japan mission BepiColombo will head for Mercury. It is named after the Italian mathematician and engineer Giuseppe Colombo, who first explained the resonance between Mercury's rotation period and its orbital period (three rotations on its own axis for every two orbits of the Sun). He also invented the Venus fly-by trajectory for Mariner 10 which allowed it to fly-by Mercury three times in 1974/5. Since Mariner 10, no other spacecraft have visited Mercury. The trip to Mercury will take three and a half years, and when it sends its data back BepiColombo should reveal much new knowledge on the history and composition of Mercury,

FIGURE 11.1 The ESA BepiColombo spacecraft at Mercury. (Courtesy ESA.)

which potentially has also a strong link to the history and formation of all the inner planets, including the Earth. Since life began soon after the formation of the Earth this may well deepen our understanding of the origins of life.

The mission will be carried out by two orbiters, one provided by ESA and the other by Japan, and potentially also a small lander. The ESA orbiter will study the surface and composition of the planet and the Japanese orbiter will study its magnetosphere, the region of space dominated by its magnetic field. The lander would be used to gauge the chemical composition and physical properties of the surface. Using a combination of solar electric propulsion and gravity assists from the Moon, Venus and Mercury itself, the BepiColombo spacecraft will brake against the Sun's powerful gravity and, once at Mercury, will use the gravity of the planet and a conventional rocket motor to ease themselves into polar orbits. All this presents considerable technological challenges.

FIGURE 11.2 A composite picture of Mercury generated by Mariner 10. (Courtesy NASA.)

In April 2009, the NASA MESSENGER spacecraft will arrive to orbit Mercury after two Venus and two Mercury fly-bys. It will carry a suite of instruments to answer basic scientific questions about the planet, such as what is the reason for Mercury's high density? And what are the compositions of its surface and thin atmosphere? What makes up the polar caps? MESSENGER will carry neutron and gamma-ray spectrometres to see if there is hydrogen in water ice at the poles. A laser altimetre will measure the polar caps' topography and thickness and answer questions about the planet's core, mantle and crust, and if the magnetic field is created by a liquid core. A magnetometre will measure the spatial and temporal structure of the magnetic field and an ultraviolet spectrometre will answer questions about the composition of the thin atmosphere. All of these instruments will have been miniaturised to meet the critical mass and volume constraints of the spacecraft.

In general terms, the knowledge we gain in the next decade from the study of Mercury will allow a far deeper insight into the formation and early history of the inner planets, including the Earth. Valuable clues as to the linkage of that early history with the origins of life can be expected.

VENUS

The planet Venus is often described as the twin of the Earth. Yet for twins the two planets could hardly be more radically different. This applies above all to their respective atmospheres. The atmosphere of Venus has a pressure 90 times that of the Earth, and consists mostly of carbon dioxide. There are droplets of sulphuric acid and other noxious gases which form extremely dense and swirling clouds, completely cutting off the surface from external view. This thick blanket creates a severe greenhouse effect and the surface temperature climbs up to 450 °C. Apart from the Moon, Venus is our closest planetary neighbour and draws twice as close to us as Mars, so many spacecraft have been sent to this planetary hothouse. In the days of the Soviet Union, the Russians sent 12 Venera spacecraft, and the NASA Magellan spacecraft provided a full global radar map of the surface. But there still persist some puzzling features of the planet. Although Venus rotates extremely slowly, taking 243 days for one rotation, huge winds sweep around the planet in times as short as four days. What drives this? Also the craters on the surface appear to be no more than 500 million years old. What happened to the older craters? The suspicion is that there is a very strong linkage between the atmosphere and the surface.

To study that linkage from the point of view of the atmosphere, ESA will send its Venus Express spacecraft to the planet in November 2005, where it will orbit at altitudes ranging from 250 km to 66 000 km.

Since it hardly seems possible that life as we know it can survive on Venus, what we learn about planetary atmospheres from Venus Express may shed some light on the early atmosphere of the Earth which, as we have seen, played a pivotal role in the emergence of life on our planet.

The US National Academy of Sciences recently recommended a Venus In-Situ Explorer (VISE) mission within the next decade to provide key measurements of the Venusian atmosphere and surface and test technologies for sample return missions in the subsequent decade. Key technological features of the mission to this inhospitable environment are an aeroshell entry into the atmosphere, passive insulation and survival, a surface drill to recover samples within one hour and a balloon ascent package to bring the sample to a survivable (for a number of days) altitude. Instrumentation is proposed to measure the

FIGURE 11.3 The surface of Venus as seen from Magellan. (Courtesy NASA.)

detailed composition of the atmosphere and its noble gas isotopic abundances, acquire descent and ascent meteorological data, measure cloud level winds, take near-infrared descent images, and measure elemental abundances and mineralogy of a core from the surface. This mission is considered essential for determining why Venus has evolved such a different environment from the Earth.

SOUTH POLE BASE ON THE MOON

The Moon is a mere three days' travel from the Earth, whereas journeys to Mars will take several months to a year. In contrast to the Zubrin 'Mars direct' approach, intermediate steps, such as manned missions to and presence on the Moon, could be extremely useful in preparation for journeys to Mars. The Moon could be a test bed for technologies to be used in such Mars missions. In addition, several of the Moon's resources could be used for *in-situ* exploitation, such as the ^{3}He (helium-3) that has been embedded in the lunar soil by the solar wind over the billions of years of its existence. ^{3}He is a valuable resource used for nuclear fusion reactors, which are considered as one of the upcoming solutions to the long-term energy demands of the Earth. Although the lunar soil contains only 13 mg/ton of ^{3}He down to a depth of about 5 metres, calculations by H. J. Balsiger have shown that the lunar resources of ^{3}He could be sufficient to cover the energy needs on Earth for several centuries (see Bibliography for Chapter 7).

However, permanently inhabited lunar bases will bring new challenges to human spaceflight, related to the radiation environments, the gravity levels, the duration of the missions and the stay time on the Moon, as well as the level of confinement and isolation that crews will need to endure. There will be immedate health issues, above all, radiation health, gravity related effects and psychological issues, any of which could become limiting factors for human adaptability and survival in these conditions.

The safety provided so far in LEO (e.g. on the ISS) will have to be greatly extended. Crew health and performance will have to be ensured during transfer flights, during lunar surface exploration, including EVAs, and upon return to Earth, as defined within the constraints of safety objectives and mass restrictions. Therefore, numerous key issues in the life sciences need to be addressed prior to the design of a lunar base.

The lunar south pole is a suitable site for a lunar base, mainly because it offers nearly permanent sunlight if located on the upper rim of a crater. Working in an area of long-lasting sunlight will significantly simplify the design of the electrical power system, leading to considerable mass and cost savings compared to a lunar base close to the equator, where 14 days of sunlight alternate with the same period of absolute darkness. An added advantage would arise if water is present in the inner part of the crater that is permanently shaded. Some observers believe that residues of water will be found in these craters as relicts from impacts by water-rich comets. However, although the Clementine missions have given some support to this idea, the occurrence or not of water in these craters is still not settled. These permanently shaded craters in the polar region are also ideal locations for astronomical activities, such as IR astronomy that requires deep cooling of the instruments. Propellant storage is also possible.

The lunar base would start with a small habitation module to which a laboratory module and all necessary scientific and operational infrastructure would be docked. If modularly built up, the different modules could be transported to the Moon by an enhanced Ariane-5 launcher or by a Space Shuttle derived vehicle. An extended lunar base at the south pole of the Moon would be composed of four modules for habitation and laboratory purposes. A crew rescue vehicle is attached, allowing an emergency return to the Earth within 24 hours.

The optimal size of the crew working in the lunar base is a key issue. On the one hand, a large crew size would offer a substantial scientific and technological return; on the other hand, each member added to the crew increases the complexity of the mission and therefore the total mass and programme cost. In the Apollo programme, a crew size of three was used; however, these missions were of short duration with a modest scientific programme and were carried out for the most part by military-skilled astronauts.

In contrast to these early lunar missions, science and technology will be the drivers for a future lunar base programme, requiring a high level of education and training of the crew in scientific and/or technological matters, such as biology, medicine, geology and/or astronomy. In addition, the following technological skills are essential for a safe and reliable operation of a lunar programme: spacecraft engineering, manufacturing technology and software engineering. Assuming that each crew member is intensively educated in at least one scientific and one technological area, a crew size of four seems to be optimal, which will be substituted every six months by supply flights from the Earth. Depending on the specific objectives of each future mission period, the crew's mix of skills can be adapted to the individual requirements of that mission. This compares to similar conclusions arrived at by Robert Zubrin for human Mars missions.

MARS

The topography of the Martian surface divides itself into four parts. The low altitude northern hemisphere with a small number of craters contrasts sharply with the multi-cratered highlands in the southern hemisphere. The equatorial region is dominated by volcanoes and great canyons like the famous Vallis Marinaris, as wide as Belgium and as long as the distance from the Urals to the Atlantic Ocean. Finally we have the polar regions with their polar ice caps. Our knowledge about these four regions of Mars has been enhanced tremendously in recent four years by the Mars Global Surveyor (MGS) mission launched in 1996, the Mars Odyssey spacecraft which started to orbit the planet in 2002 and Mass Express since 2004.

Although these spacecraft have revealed to us a more complex story about Mars than we believed a decade or so ago, they also reveal and confirm some simple facts. The chief among them is that the Martian atmosphere and surface are dominated by dust and wind. The dust is everywhere and the wind blows everywhere. When the Mariner 9 spacecraft arrived in Mars orbit in 1971, it had to wait weeks before a global dust storm abated and it could take the first long series of photographs which gave us the first synoptic assessment of the planet. It was these Mariner 9 pictures which revealed in detail the planet-wide cratered surface but also the huge outflow channels and apparent ancient river beds which so captured the public imagination. The story of liquid water on Mars has many nuances. If we speak of water in general, we now know or at least infer from the epithermal neutron spectrometry of Mars Odyssey that aside from equatorial regions there is considerable water in all regions, mostly in the form of ice and ice mixed with dust.

Other facts about Mars have also recently been discovered. Although the planet does not presently have a magnetic field like the Earth does, there are a large number of areas of the surface where the iron-rich rocks are magnetised. This indicates that the planet had a magnetic field in the past when the rocks solidified. But where it came from and where it went to remain unknown. Also new facts of geology have emerged. Whereas the rocks of the highland and highly cratered southern hemisphere are mainly igneous basalt, the northern hemisphere is mostly andesite, which is also a volcanic rock but more complex than basalt. The reason for this differentiation is unknown. There is a suggestion that the higher, more cratered southern hemisphere is older and that the lower northern plains are smoother because of erosion by liquid water in the form of oceans or lake beds. There are even hints of shorelines around the southern highlands. In the northern hemisphere there appear to be hidden craters, as if they had been hidden by erosion processes such as those made by flowing liquid water.

The second simple fact about Mars that has been revealed by MGS is that almost all of the Martian surface is covered by layered deposits. There are layers evident wherever the crust is exposed, such as the walls of canyons and valleys, on mesas and on the side walls of craters. And the layers are all different from one another in visual appearance and in thickness. Apparently, a complex series of cratering, deposition and erosion events have taken place throughout the planet's history. It seems fairly clear that volcanism has played a far

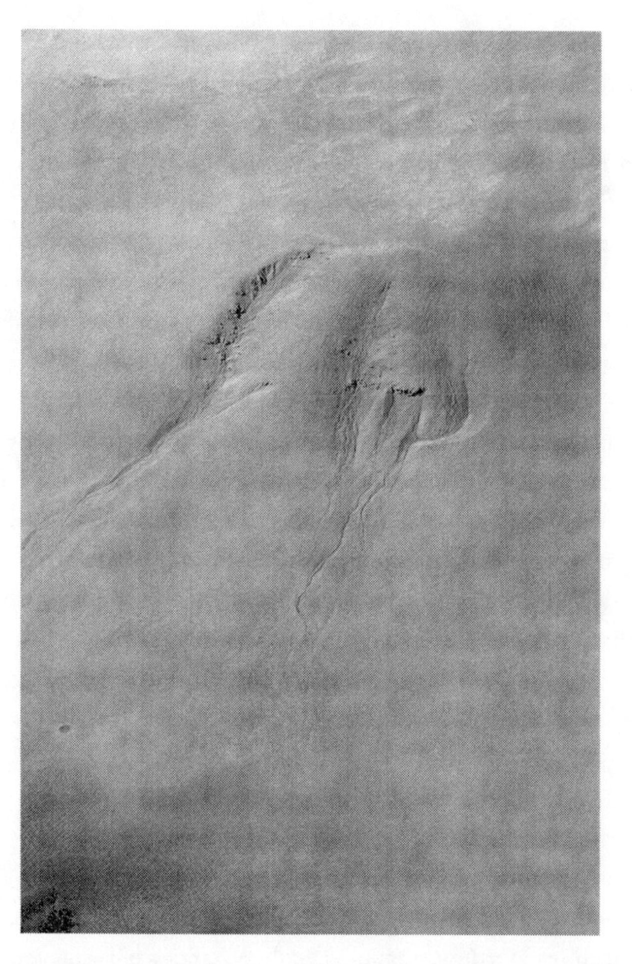

FIGURE 11.4 Traces of water trickling down slopes, as imaged by MGS. (Courtesy NASA.)

more minor role in this process than the effects of impacts, a conclusion that was also reached in the 1970s regarding the Moon. In this instance we have learned that Mars is much more complex than we first imagined. The differentiated and complex nature of the great planets and their moons as revealed by the Pioneer and Voyager spacecraft in the 1970s surprised everyone. Perhaps the lesson here is that the more closely we look at objects with a 4 to 5 billion year age, the more we should expect complexity, nuance, variety and surprise.

If comets deliver oceans to the Earth (see Chapter 2) then they should deliver similar amounts of water to other planets such as Mars. There is ample evidence for ancient water on Mars, huge outflows, gullies, river beds, flooded craters and, more recently, MGS pictures of water trickling down slopes in the recent past (~1000 years ago).

But where did the Mars water go to? Certainly there is still some water vapour in the thin atmosphere and there are considerable quantities of water in the form of ice at the polar caps. There may also be water in the form of permafrost just below the surface. Further down there may even be liquid water melted by geothermal heat. Upcoming missions will carry ground penetrating radar to search for water many kilometres below the surface. This pursuit of water is called the 'follow the water' strategy in the search for life. And when humans go to Mars they will be doing *in-situ* geology in craters which have been filled with water and may have sedimentary deposits with fossils in them. Ultimately, we should be able to settle once and for all the question of whether or not water on Mars, billions of aeons ago, led to the appearance of life on the now apparently dead surface of the red planet.

The question of water on Mars, where it came from and where it went to, is one of the major scientific issues of our time. When Mariner 9 made the first detailed images of the Martian surface, showing not only the now famous cratered surface of the planet but also the huge outflow channels, gullies, and what appeared to be ancient river beds, the world stood in awe at what was clearly the relic of a past geology dominated by flowing liquid water. This issue continues to occupy the best minds of planetary science. But recently a more intriguing light has been cast on the Martian water story that hints that it may not be just an ancient water story after all.

By far the most stunning discovery is that of gullies indicating flowing water as recently as a few thousand years ago, which in geological terms is essentially just yesterday. These gullies first showed up in the ground-breaking images from NASA's Mars Global Surveyor spacecraft. MGS arrived at Mars in September 1997. The gullies usually occur in clusters on the slopes of hills or crater walls at Martian latitudes between 30 degrees and 70 degrees in both the northern and southern hemispheres. They usually start a few hundred metres from the top of the slope, can be up to a few hundred metres in width, up to several kilometres in length and and several tens of metres deep. Typical gully widths are 20 metres, 500 metres in length and about 10 metres deep. Although other flowing fluids as causes of these gullies have been proposed, such as liquid or gaseous CO_2 or high concentration brines, the overwhelming consensus is that they were/are produced by flowing liquid water.

But the most surprising aspect of these gullies is that they appear to be only a few thousands of years old and so the inescapable conclusion has to be that liquid water must have flowed on Mars in the very recent geological past. The source of this water is still a matter of discussion amongst researchers. One theory is that it is surface run-off from subsurface aquifers, but the mid-latitudes are cold and liquid water from this source seems unlikely here. Another theory proposes that it is from melting ice near the surface that happens when warming conditions occur. A more recent proposal by P. R. Christensen is that the water is from melting snow that has been transported to the mid-latitudes from the poles in periods of high obliquity during the last 100 000 to 1 million years. In this theory, the water produced by this snow can create these gullies within about 5000 years.

The significance of the latter theory is that it provides a mechanism whereby a stable liquid water supply can be generated in snowy deposits within a few centimetres of the surface with a geologically short regeneration time, providing a potential mechanism and environment for life to survive over geologically significant periods of time.

Another curious feature discovered by MGS is that there appears to be a considerable degree of erosion of many of the surface features of Mars at the present time, almost as if global climate change is happening as we look on. At the planet's south pole, alternating layers of dust and ice are disappearing. Using pictures taken in 1999 and 2001, observers have calculated an erosion rate of 3 metres per year. It seems

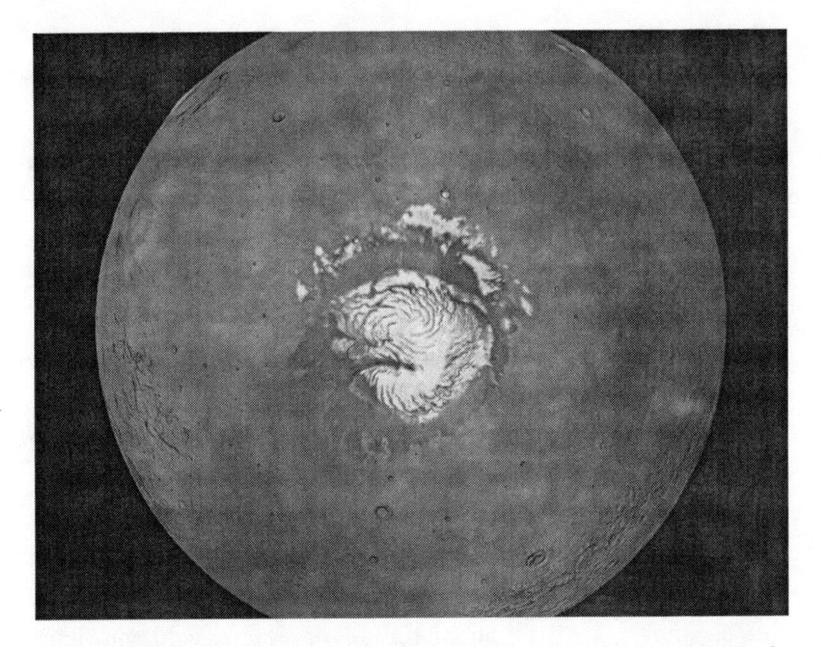

FIGURE 11.5 The north pole of Mars as imaged by Viking in 1976 (see also Plates 18 and 19). (Courtesy NASA.)

as if the carbon dioxide ice is melting and vaporising at a rate that could lead to erosion of the permanent ice cap in less than a century. This could also lead to a thickening of the Martian atmosphere. There appears to be a real time global climate change taking place on Mars. Clearly Mars has still many surprises to unveil.

At the present time the interest in Mars is at an all-time high. An armada of spacecraft have and are being sent to the red planet in pursuit of its secrets. After the tremendous success of MGS and the Pathfinder surface robot, NASA followed this with the two spectacular failures of Mars Polar Lander and and Mars Climate Orbiter in 1999. However, in 2001 NASA had an interesting success with the Mars Odyssey spacecraft. Onboard this orbiter is a series of neutron detectors which can pick up evidence of the presence of water ice. In Figure 11.5 we see the image of the north pole of Mars as seen by one of the Viking spacecraft in 1976. The Odyssey spacecraft has imaged the same pole in in winter and then in summer 2001 with its neutron imagers. In Plate 18 is shown the situation in winter when the pole is covered with carbon dioxide ice and the amount of water ice 'visible' is limited. In summer, the carbon dioxide ice melts to reveal the degree of water ice that was below (Plate 19). There seems little doubt that with these quantities of water ice on Mars there must be at least some quantity of liquid water associated with them. And according to the 'follow the water' philosophy, it is in areas like these that we should look for water.

MARS EXPRESS

ESA's Mars Express mission carried a Mars orbiter and a lander called Beagle 2, named after the ship in which Charles Darwin sailed around the world collecting evidence for what he would later publish as the theory of evolution. The Mars Express spacecraft weighed in at just over a ton with a 113 kg orbiter payload and the 60 kg Beagle 2 lander. A comprehensive set of instruments on board the orbiter brings much new science to the observation of Mars:

Surface science

- a *high resolution stereo camera* images the planet in 3D colour with a resolution of 10 metres. In some cases a 2-metre resolution is possible, allowing many other scientific measurements.
- a *subsurface sounding radar/altimetre* uses low frequency radio waves to detect features below the surface, such as the interfaces between different materials, including ice and water. Any subsurface water will be detected by this.
- a *visible and infrared mineralogical mapping spectrometre* uses visible and infrared light reflected from the surface to measure the mineralogical composition of rocks, such as their content of iron, water, sulphates and carbonates.

Atmospheric science

- an *energetic neutral ions analyser* investigates how atoms of oxygen and hydrogen in the upper atmosphere interact with the solar wind, which is thought to blow away Mars's atmosphere. Unlike the Earth, Mars at present has no magnetic field which, as in the case of the Earth, would deflect the solar wind thereby protecting the atmosphere. This research tells us how the Martian atmosphere evolved over billions of years.

- an *ultraviolet and infrared atmospheric spectrometre* uses ultraviolet light to study ozone concentrations, and infrared radiation to study water vapour concentrations. Seasonal variations of these key gases in the Martian atmosphere can be gauged.
- a *planetary Fourier spectrometre* measures the vertical profile of temperature and pressure of the major component of the atmospheric carbon dioxide. It will also measure concentrations of less abundant components such as methane, water, carbon monoxide and formaldehyde. On Earth, some of these minor components have key links to the emergence of life.
- a *Mars radio science experiment* will, by measuring the variation in the radio signals between the Mars Express orbiter and the surface, give information about the atmosphere, ionosphere, surface and interior.

The Beagle 2 lander consisted of a set of solar panels/batteries to provide power, a robotic arm carrying most of the experiments and a gas analysis package. The robotic arm could stretch and rotate and carried two stereo cameras to image the landscape. These cameras could also focus on nearby rocks and soil samples to pick out interesting candidates. Using a grinder to clean the surface allows analysis instruments such as a microscope to determine whether a rock is sedimentary or volcanic, a Mössbauer spectrometre to measure the iron content of the rock and an X-ray spectrometre to determine the age of the rock using the potassium-40 isotope.

The robotic arm was to have been used to bring the samples identified as most interesting from the point of view of a search for life back to the analysis package located within the main Beagle 2 structure. The package housed 12 ovens in which samples are heated in the presence of oxygen. The resulting carbon dioxide, which is produced

FIGURE 11.6 The ill-fated Beagle 2, as it would have deployed on the Martian surface. (Courtesy ESA.)

at different temperatures, is measured in a mass spectrometre where the carbon-12/carbon-13 ratio – a clue to the existence or not of life – is determined. Other gases could also have been measured, including atmospheric methane, a potentially key indicator of the presence of extant life on Mars.

The Mars Express orbiter will operate up to 1 Martian year (678 Earth days). A further Martian year is possible. The Beagle 2 lander would have operated for six Earth months had it not been lost.

EXOMARS

ESA's EXOMARS mission based on the robotic search for life on Mars will be launched in 2009. It will be the first Mars mission dedicated primarily to exobiology. It will carry a rover and an integrated exobiology science package on the rover capable of drilling 1 to 2 metres below the presumed oxidising layer, where traces of extinct life may have survived and where there might just be signs of extant life. The science package is named 'PASTEUR' after the great French scientist who did so much to enhance our understanding of the phenomen of life. It will include most of the analysis instruments listed in the

ESA science team report ESA SP-1231 referred to in Chapter 4, including:

- A gas chromatograph/mass spectrometre
- A Raman spectrometre
- A Mössbauer spectrometre
- An imaging microscope

In addition, the package will probably include an electromagnetic subsurface water detector and a microchip-based multiorganics detector.

The NASA robotic missions to 2015

Mars Exploration Rover missions in 2003–2004

The two rover missions called Spirit and Opportunity were launched in summer 2003. They landed on two separate sites on Mars and had as objectives the seeking out and characterisation of soils that would reveal facts about past water on Mars. As we have seen, the presence of liquid water is fundamental to the existence and survival of life, and Mars displays extensive evidence of flowing and stationary bodies of water in its past. The first task of the rovers at each site was taking panoramic images of the surrounding terrain in order to pick out promising or interesting targets for further, closer examination. Then the rover manoeuvred to those targets so that onboard analysis instruments could start to carry out these analyses. The analysis instruments that performed this were (i) a thermal emission spectrometre for finding out how Martian rocks were formed, (ii) a Mössbauer spectrometre for analysing iron-bearing rocks, (iii) an alpha proton X-ray spectrometre (APXS) for measuring the abundances of the elements in rocks, (iv) magnets for picking up magnetic particles and analysing them using the APXS, (v) a microscope for close-up high resolution images of soils and rocks, and finally (vi) a rock abrasion tool to clean off any weathering of the rocks prior to their analyses by these instruments. (See Plate 21.)

The rovers moved from site to site carrying out these analyses during their mission lifetimes. They were able to travel up to 40 metres a day, up to a maximum of 1 kilometre. The panoramic stereoscopic imager is mounted on a 1.5-metre mast providing a 360 degree, almost humanlike surveying of the landscape. The rock analysis instruments are mounted on a robotic arm allowing them to be brought directly up against the soil and rock samples.

The landing sites had been chosen after extensive consultations and discussions with the scientific community. Gusev Crater is a very interesting feature that seems to have a high probability of having had water in it. There is a large channel that enters Gusev from the south and this may have created a lake in the crater. The other landing site was Meridiani Planum and its choice was facilitated by data returned by Mars Global Surveyor. This site has extensive deposits of haematite, which is an iron bearing mineral (iron oxide) and on Earth is usually associated with minerals being processed by water and heat, i.e. a hydrothermal environment. The choice of these two landing sites from an initial list of over 100 was driven by the 'follow the water strategy' that dictates that search-for-life related investigations should go where there is a high probability that there has been liquid water. The landing sites were near the equator in order to make use of the maximised solar energy available there.

The 2005 Mars Reconnaisance Orbiter

This spacecraft will have a powerful telescopic camera and be able to image objects the size of a small table, creating an enormous database of geographical detail. It will return extensive data on surface minerals, particularly water-related minerals, and also use ground-penetrating radar to search for ice and even liquid water hidden beneath the surface. It will reveal key surface and subsurface data in preparation for future lander missions.

Robotic missions post 2005

There are mission opportunities to Mars every 26 months. In 2007, NASA plans to start a series of small Scout missions. The first of these

FIGURE 11.7 Spirit on Mars. (Courtesy NASA.)

will be 'Phoenix', from the University of Arizona, that will go to the water ice-rich northern polar region and with a robotic arm dig into the arctic surface looking for suitable habitats for micro-organisms. Further Scout mission plans include a rocket powered aeroplane on Mars, an orbiter to look at trace gases in the atmosphere that might indicate biological processes, and an atmospheric sample return mission.

In 2009, a major step forward is expected with the launch of the Mars Science Laboratory. This will be a long-duration, long-range roving science laboratory to look at carbon chemistry on Mars. It will demonstrate smart capabilities for landing at what may be biologically interesting but logistically difficult sites. It will pave the way

FIGURE 11.8 The Mars Reconnaissance Orbiter. (Courtesy NASA.)

for the very technologically challenging sample return missions that will have to collect samples on Mars, load them into a suitable spacecraft, bring this to a safe rendezvous with a Mars orbiter and return this complex safely to Earth. The plans for this sample return presently foresee this taking place in 2014. This return capability will of course be mandatory for human missions to Mars.

Human Mars missions

From a creative point of view, the plan of Robert Zubrin for the first and subsequent human missions to Mars and the way they will lead ultimately to the human colonisation of Mars has, at the present time, no serious contenders. A NASA plan for Mars developed in response to President George Bush's 1989 call for a Space Exploration Initiative named the '90-Day Report', resulted in a cost estimate of $450 billion and called for huge infrastructure elements like orbiting spacecrafts, advanced propulsion technologies and huge interplanetary cruisers. Frustrated and appalled, Zubrin, an engineer at the Martin Marietta company in the USA, Ben Clark and David Baker, another engineer of the company, developed the 'Mars Direct Plan'. It became the basis of the ideas of NASA's human Mars mission planning as reflected in the NASA Human Mars Reference Mission, which to this day represents the basic thinking about those missions.

Tellingly, in his book *The Case for Mars*, Zubrin emphasises the difference between two opposing means of exploration, i.e. the 'dog sledge' approach used by Amundsen in his successful traverse of the North West Passage, as opposed to the huge British Admiralty expedition led by Sir John Franklin in 1845, using two 300-ton ships replete with huge quantities of provisions and luxuries. Within months, the Franklin expedition had vanished without trace, eaten up by the Arctic wilderness – an expedition completely suited to the grandiose notions of the British Admiralty's understanding of exploration of the wilderness, but completely unsuited to the need for flexibility and learning from the land as you go and similar strategies as practised by Amundsen.

The key elements of Zubrin's approach are:

- The use of *in-situ* resources on Mars to support the stay of humans and their safe return to Earth
- The use of existing technological capabilities

Concerning the first point, Zubrin's fundamental insight is that a major burden on the whole mission is the need to carry, on the outward journey to Mars, the fuel needed for the return journey to Earth. In this case the severe consequences of the 'Rocket Equation' referred to in Chapter 10 have their full impact. If, however, one can find the fuel for the return journey at Mars itself, then the equation changes dramatically. Now, the major freely available resource on Mars is its atmosphere, which is almost entirely (95 %) composed of carbon dioxide CO_2. According to a process known since the nineteenth century, CO_2 can be reacted with hydrogen H_2 to produce methane CH_4 and water H_2O according to:

$$CO_2 + 4\,H_2 \rightarrow CH_4 + 2\,H_2O$$

This is called the Sabatier reaction, as previously described, after the French chemist of the nineteenth century who studied it. It is exothermic, i.e. releases energy, and is highly (99 %) efficient. The methane produced in this reaction is then liquefied by the cryogenic hydrogen and stored as the future fuel. Then the water is broken down by electrolysis:

$$2\,H_2O \rightarrow 2\,H_2 + O_2$$

where the oxygen is refrigerated and stored as the future oxidising agent for the return to Earth and the hydrogen is recycled back to the Sabatier process.

It cannot be emphasised enough that this is just elementary school chemistry. There are no fundamental scientific or technological impediments to the practical implementation of this on the Martian surface. The only burden then on the outward journey from Earth is to bring the liquid hydrogen to Mars to feed the Sabatier reaction. This dramatically changes the picture. Apart from the fuel consumed in launching Mars vehicles into Earth orbit and then on their way to Mars – exactly like the outward journeys of the Apollo vehicles to the Moon – the Mars vehicle only has to bring 6 tons of liquid hydrogen to Mars. In combination with the carbon dioxide in the Martian atmosphere, a total of 108 tons of methane and oxygen will be manufactured by the Sabatier reactor carried out to Mars by the first uncrewed Earth Return Vehicle. Only when flight controllers and engineers on Earth have established that the ERV is fuelled up and capable of a safe return to Earth will the HAB module carrying a crew of four be launched to Mars.

Now the key issue here is that due to the vastly reduced burden of fuel on the outward journey, existing propulsion technology, namely Saturn V size rockets, each employing four shuttle main engines and two shuttle solid boosters, can be used. Two of these launches – one for the uncrewed ERV and one for the crewed HAB module – would be required for each mission. With two launches every two years, an almost continuous human presence on Mars could be achieved.

The crew in the Zubrin scenario

In the Zubrin scenario considerable thought was given to the number and competences of the crew. Although he points out that in other studies an 'as many as possible' approach was fairly common, his belief is that to minimise the demand on infrastructure and support logistics, and bearing in mind that risks are being taken, from a logistical and moral point of view the optimum crew is the minimum crew. Now since the major threat to the safety of the mission and the crew is mechanical and/or electrical failure of key systems and sub-systems the most vital member is therefore the *flight engineer*. Because it is so vital a function, Zubrin recommends that two crew members serve in this function. Next in importance, and recognising that the mission's objectives are the *scientific exploration* of Mars, Zubrin advocates two *field scientists,* one a geologist to explore the geology of Mars and one a biogeochemist to carry out the search-for-life work. For Zubrin then, that's the minimum crew. Other functions, such as the role of commander who takes decisions, pilot and medical care, can be assigned to individual crew members on the basis of individual cross-training. Zubrin's argument has little place for one-off functions such as manager (commander), pilot (it is easier to train a geologist as a pilot than the reverse) or expedition doctor (he cites examples of earlier explorations where the explorers themselves can provide emergency medical aid).

Whereas the overall technical feasibility of human missions to Mars had already been proven half a century ago, as described in his book *The Mars Project* by Wernher von Braun (1953), human missions to our sister planet Mars represent long-duration undertakings with trajectory times of several months to a year. Preliminary plans developed in the USA as well as in Europe foresee that a first human mission to Mars – which can probably only be undertaken in an international joint effort – may happen before the middle of this century; the date is slowly moving backwards, from 2018 – a date used in the HUMEX study – to between 2020 and 2030 – dates now tentatively envisaged by the space agencies NASA and ESA. From the different mission scenarios under discussion, two are considered as the most likely: (i) a 1000-day mission with more than 500 days' stay on the surface of Mars, and (ii) a 500-day mission with only a 30-day stay on Mars. For both types of mission, a suitable launch window opens about every 26 months. The launch window is dependent on the relative position of the Earth and Mars during their orbits around the Sun. We will concentrate on the 1000-day Mars mission because (i) it is technically more feasible than the short-term mission since it requires less energy, and (ii) it allows sufficient time on the surface of Mars for scientific on-site investigations. In a 1000-day mission profile, the flight to Mars lasts typically 200 to 300 days. After arrival at Mars, the crew will stay on Mars for 400 to 500 days, waiting for the next low energy launch window to open for the return to Earth. The use of a low energy transfer trajectory is essential to significantly reduce the spacecraft's mass and therefore the total programme costs. The return flight again lasts 200 to 300 days, depending on the selected launch date. Although Zubrin makes a strong case for a crew of only four, others have argued that a larger crew is needed to ensure mission safety and success, especially in the case of illness occurring unexpectedly. Some arguments in this direction conclude with an optimum crew size of six.

Because a mission to Mars should be science-driven, the crew should be skilled in scientific areas (e.g. biology, medicine, psychology, geology, atmospheric research, meteorology, and astronomy), in addition to technological skills (commander, spacecraft engineering, manufacturing technology, navigation, communication, software engineering) that are required for the success of the mission. Assuming that each crew member is intensively educated in one scientific and one technological area, this leads to a crew size of six members. Whereas for a lunar mission, the illness of one crew member does not directly jeopardise mission safety thanks to the short transportation and communication times, this will be different for Mars missions: the serious illness of one crew member (e.g. the medical doctor) could jeopardise the whole mission due to the very long communication links (up to 45 minutes bi-directional) and transportation distances of several hundred million kilometres. Therefore, for mission success, some crew members should have a third back-up education in crucial skills like medicine, software and spacecraft engineering. This means that some or all crew members should be educated in up to three different scientific and/or technological areas.

At arrival in Mars orbit, the crew splits: two crew members remain in Mars orbit and four crew members descend to the Martian surface, where they stay and work for 525 days (see Table 6). This splitting of the crew is comparable to the Apollo programme, with one astronaut staying in lunar orbit whilst two astronauts descended to the lunar surface. The main reason for two crew members remaining

TABLE 6 The main crew activities during a 6-crew, 1000-day Mars mission.

Mission phase	Earth-Mars transfer*	Stay on Mars	Mars orbit	Mars-Earth transfer**	Total mission
Duration (days)	225	525	525	204	954
Crew number	6	4	2	6	6
EVA number***	0	175	0	0	175

*including 7 days in LEO and 14 days in Mars orbit

**including 14 days in Mars orbit

***each EVA includes 2 astronauts for 8 hours maximum

in Mars orbit is safety. The astronauts in Mars orbit have to monitor and control the parent ship to ensure safe return to the Earth at the end of the mission. In view of the accidents that occurred on the MIR space station and some technology problems with the first ISS modules, it is clear that the parent ship can only be kept in a good workable condition for more than 500 days with human presence and support. The two astronauts in Mars orbit can also provide significant support to their colleagues on the Martian surface, e.g. by remote sensing activities and by providing communication links to the Earth.

In conclusion, we can say that there are really no technological impediments to launching a human Mars exploration programme today similar to President Kennedy's commitment of one to the Moon in 1962. Certainly there are areas where technical issues need to be addressed, such as the closing of the cycle of the Sabatier process and the life support systems, the proof of rendezvous and docking in Mars orbit, the re-entry and landing on the Earth after the return from Mars, and finally the proof that humans have a good chance of surviving both physiologically and psychologically the 1000-day journey. But none of these is likely to be a show-stopper. They are all essentially variants of problems that humans have shown themselves eminently capable of solving in the past – given the resources and the commitment. And these are in turn predicated on two supremely human faculties, that is, imagination and the will to succeed, which we will deal with in Chapter 13.

THE GREAT PLANETS AND THEIR MOONS

Jupiter and its moons

The NASA Jupiter Icy Moons Orbiter (JIMO) will be a ground-breaking mission to the icy moons of Jupiter: Europa, Ganymede and Callisto. In combination with Project Prometheus to develop nuclear powered electric propulsion, it will pave the way by NASA to explore the great planets and their moons and lead the way to exploration of this important part of the Solar System. It is already almost certain that Europa harbours a liquid water ocean beneath its surface, but the icy surfaces of Ganymede and Callisto may also do so. The route to these icy moons has already been pioneered by NASA's Galileo probe, which found evidence for Europa's ocean and partial evidence for similar oceans on Ganymede and Callisto.

NASA's objectives for JIMO include:

- Determine the thickness of the ice layers and identify future landing sites. Map where organic compounds and other substances of biological interest lie on the surface.
- Investigate the origin and evolution of these moons, including the nature of their interior structures, surface composition and surface features, as this relates to their evolutionary history and how this might add to our understanding of the origin and evolution of the Earth.
- Investigate the radiation environment of Europa, Ganymede and Callisto, all of which reside within the influence of the powerful magnetic field of Jupiter, which in turn affects the charged particle impact on these moons and may play a central role in whether life can originate, evolve and survive on these bodies.

The measurement to be made on the icy moons will be primarily radar probes to measure the thickness of the ice. There will also be laser mapping of the surface elevations. The mission will carry cameras for photographic mapping of the moons' surfaces, as well as infrared imaging, a magnetometre for measuring magnetic fields and instruments

FIGURE 11.9 The Galilean moons of Jupiter: Io, Europa, Ganymede and Callisto. (Courtesy Lunar and Planetary Institute.)

to measure the radiation patterns of charged particles, atoms and the dust around and on the surfaces of these bodies.

The first proposed date for launch of this mission is 2011. It would be the first space science mission of NASA to employ the nuclear fission/electric propulsion technology. Project Prometheus has the task of demonstrating the safe and reliable operation of this type of technology in deep space missions.

Europa

Of the four Jovian moons discovered by Galileo in 1610, Europa is the second in distance from Jupiter and is the smallest. It is approximately the same size as our Earth's Moon, with a diametre of about 3100 km. Its orbit around Jupiter is in 'resonance' with the orbits of its fellow Galilean moons Ganymede and Io, and this makes its orbit quite eccentric (elliptical). As it orbits Jupiter in this eccentric orbit, the varying gravitational pull causes tidal stresses that heat both the surface and the interior of Europa. In the 1970s, and before any spacecraft from Earth ventured near it, ground-based astronomy identified the characteristic infrared signatures of water ice on Europa's surface. When the Voyager and Galileo spacecraft took close-up images of Europa's surface the remarkable similarity with ice features on Earth's polar oceans was revealed. The ice appears to be broken up into plates which move and even twist around, as if they have been floating on some sort of liquid or slushy substrate – the famous presumed Europan ocean.

Although the heat from the Sun keeps the Earth's oceans at a temperature that makes water liquid, this cannot be the case for Europa. It is too far away from the Sun for that. But the heat from the tidal stresses described above is just right to keep the water below Europa's icy surface liquefied, or at least slushy.

There is another striking feature of the surface of Europa. There are very few obvious craters. Some crater features can be seen below the ice but hardly anything on the ice. In general, solid planetary surfaces in the Solar System which are not strongly weathered by an atmosphere display extensive cratering. Mercury, the Earth's Moon,

Mars and asteroids all show the evidence of continual bombardment. This tells us that the icy surface of Europa is probably both young and in constant motion.

But the most interesting feature of all is a veined network of reddish-brown filaments crossing and criss-crossing the icy surface. Many of the filaments have lighter coloured lines along their centres. One theory suggests that the veins are fissures in the ice through which material from below is exuded. The veins themselves are tens of kilometres wide and remain mysterious in nature. If they represent material exuding from below, and there is a water ocean below with life present in it, then presumably traces of life could be found at the surface of Europa in or near these fissures. And so it may not be necessary, when searching for life on Europa, to penetrate the surface ice – the evidence may be in the red/brown fissures on the surface.

Thera (left dark, reddish region in Figure 11.10) is about 70 kilometres wide by 85 kilometres high and appears to lie slightly below the level of the surrounding plains. Some bright icy plates which are observed inside appear to be dislodged from the edges of the chaos region. The curved fractures along its boundaries suggest that collapse

FIGURE 11.10 Thera and Thrace are two dark, reddish regions of enigmatic terrain that disrupt the older icy ridged plains on Jupiter's moon Europa, as imaged by NASA's Galileo spacecraft. (Courtesy NASA.)

may have been involved in Thera's formation. In contrast, Thrace (right) is longer, shows a hummocky texture, and appears to stand level with or slightly above the older surrounding bright plains. Thrace abuts the grey band Libya Linea to the south and appears to darken Libya. One model for the formation of these and other chaos regions on Europa is a complete melt-through of Europa's icy shell from an ocean below. Another model is that warm ice welled up from below and caused partial melting and disruption of the surface.

Other evidence for a watery ocean below the icy surface was found by NASA's Galileo spacecraft in early 2000. The magnetometre instrument aboard Galileo found a varying change in the direction of Europa's magnetic field. This type of fluctuation is what would be expected in an electrically conducting medium such as a salt-rich ocean lying underneath the surface. This is because the magnetic field of the parent planet

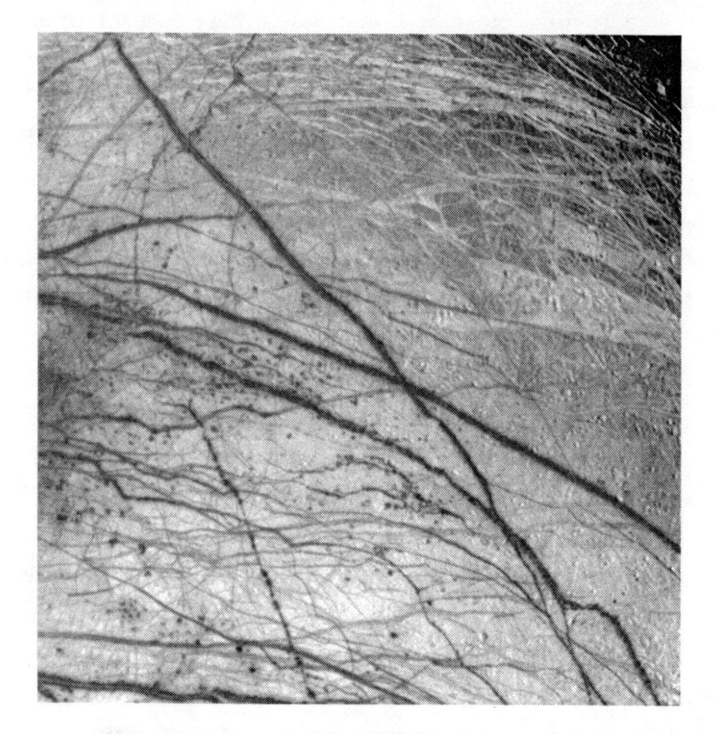

FIGURE 11.11 An image of the Minos Linea region of Europa showing the veined network of filaments. (Courtesy NASA.)

Jupiter varies over a cycle of about 6 hours. This induces electrical currents in the conducting layer (the presumed salty ocean) that in turn produce a magnetic field of Europa itself. This magnetic field has its magnetic poles at or near Europa's equator and the direction of the field reverses itself every 6 hours. Interestingly, another moon of Jupiter, Callisto, also displays this type of variable magnetic field, possibly with a subsurface salty ocean with a depth of 10 km.

Liquid water is indeed a prerequisite for life, but life also requires oxygen, carbon, sulphur, phosphoros and nitrogen. Is it possible that these elements could find their way to Europa's icy surface? The answer appears to be yes, and the source is comets. Based on what is known about the amount of organic molecules, including amino acids, and the number of impacts of comets with planets and moons in the Solar System, and even though Europa has a weak gravitation due to its small size, it turns out that comets could have brought enough of these biogenic elements to support an amount of biomass equal to about 1 % of the biomass of procaryotic (simple early life) cells on Earth.

The third important ingredient needed for life is energy. There must be a source of free energy within the medium where the lifeform exists in order to drive the metabolic processes characteristic of life. On Earth, this energy is derived from sunlight by the process of photosynthesis. But there must be a sufficient intensity of sunlight to provide the energy. In a Europan ocean below many kilometres of ice, where ice only transmits light a few metres, there can hardly be any energy available for photosynthesis. But there are other possible sources of energy, and it should be recalled that the Earth probably had a biosphere supporting life before photosynthesis evolved.

Chief among the candidates are hydrothermal vents, similar to those found on the seabed of Earth's oceans (see Plate 6). These would necessarily need to rely on the presence of volcanic activity within the core of Europa which is by no means evident, either on observational or theoretical grounds.

What would Europan organisms be like? They would probably be micro-organisms which rely on chemical reactions to provide the energy for metabolism (chemosynthesis) rather than light (photosynthesis), since the Europan ocean will certainly be a dark and murky place. They might resemble the hyperthermophiles on Earth, which thrive on the heat near hydrothermal vents on the ocean floor.

The question now arises, what type of missions can we conceive to search for life on moons such as Europa, Callisto and possibly Ganymede? These could be:

- Robotic surface exploration mission/s
- Robotic probes into the ice to reach the subsurface ocean
- Human surface landing with Antarctic type bore to perform drilling

1. *Robotic surface exploration mission/s*
   This type of mission is entirely feasible for launch within the next 10 to 15 years. The journey time to the Jovian system is only about 3 years and a spacecraft could be used to land on the icy surface, or at least perform an orbiter mission such as the JIMO mission described above. It can use radar, magnetometry, gravity sensing and altimetry measurements to tell us about the subsurface of the moon. The first question will be – is there a liquid water ocean below the solid ice surface? Using a laser altimetre, the tidal deformation of Europa's surface described above would reveal all: for a fully solid shell of ice, only 1 m deformations are expected, whereas if there is a liquid ocean below, then deformations up to 30 m are expected. In addition, a radar sounder would probably be able to detect any water/ice boundary. It is also possible to detect from orbit organic and exudant substances on the surface of Europa using infrared spectroscopy. Surface lander missions would almost certainly target surface features such as filaments and chaotic regions described above to see if they are indeed exuded from an ocean below and to see if they contained any organics or traces of life.
2. *Robotic drilling missions*
   If it turns out that the surface features of Europa, such as the veined filaments, are not exuded from a subsurface ocean, then the next

missions would be landers on the ice with probes to melt their way into the ice and down to the ocean below. This melting would require significant amounts of power and this would almost certainly have to be provided by a small nuclear reactor. Depending on the thickness of the ice and our ability to control and direct the probe remotely, we might find ourselves at the limits of what is possible with robots, however smart they have become. Ultimately, we may have to envisage human missions to Europa to investigate its ocean on the spot and reap whatever benefits present themselves.

3. *Human drilling missions*
   Imagine it is the year 2040. There are human colonies on Mars and routine traffic between the Earth and Mars. Nuclear thermal propulsion for human spaceflight is coming on line in a reliable way. Human missions to the moons of the Great Planets are now possible. Due to the distances involved, robotic missions to the Jovian moons have only met with partial success. On Earth, a huge biotechnology industry has grown up based on the application of genetic sequences from extremophiles found on Mars in the late 2020s. A vast array of new drugs have been developed as a consequence of these discoveries and the average life expectancy in the developed countries has been increased by between 10 and 20 years. The prospect of finding new life forms and new genetic sequences with huge new pharmaceutical potential in Europa's ocean beckons. There is now no real technical challenge. All it needs is commitment and investment. Decisions are taken. An Antarctic type base is planned for Europa with a drill technology based on the by now famous Russian drill that found Lake Voskhod beneath the Antarctic ice in the late 1990s.

This may seem the stuff of fantasy. But everything stated here is neither fantastical nor improbable. Many biotechnology processes are now in place as the result of research using extremophiles found on Earth. Developments in the medical field and particularly in pharmacology have increased life expectancy in the developed world dramatically since the end of the Second World War. Nuclear thermal

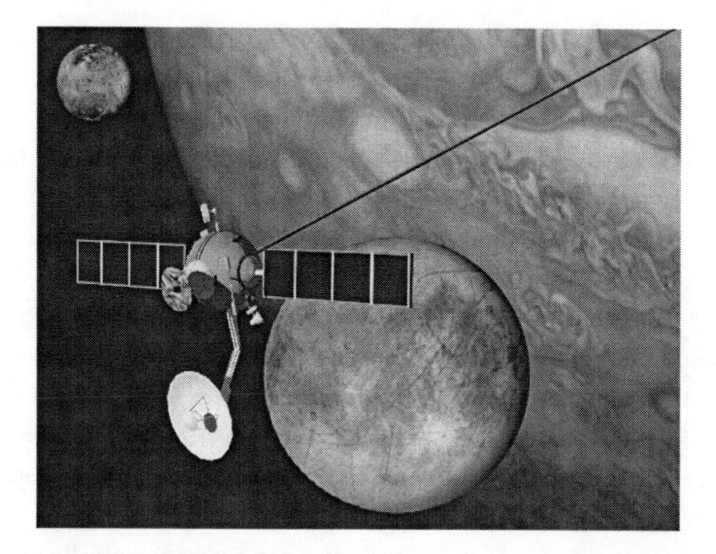

FIGURE 11.12 An Electrodynamic Tether Probe approaches Europa. (Courtesy NASA.)

propulsion has been demonstrated in the Nevada desert in the 1960s. Europa clearly beckons.

Ganymede and Callisto

Ganymede is the largest satellite in the Solar System and is about the same size as Mercury, but its low density indicates that its composition is about 60 % water and 40 % rock. The surface appears to consist of a dark, heavily cratered terrain, which is assumed to be old, and brighter, younger regions with grooves and ridges which can be over 100 kilometres in length and tens of kilometres wide. There are indications that Ganymede and its smaller sister moon Callisto may have subsurface oceans similar to Europa. If this turns out to be the case, then attention similar to the attention likely to be directed at Europa for the reasons given above will also fall on these bodies.

SATURN AND ITS MOONS

In 1980, NASA's Voyager 1 spacecraft flew by Titan, Saturn's largest moon, and discovered that it had an atmosphere thick with nitrogen and carbon compounds and, in particular, methane. As such, it may resemble the Earth's early atmosphere if indeed that was a reducing (no oxygen) one, as supposed by Oparin and the Miller-Urey experiment (but see Chapter 2 where the modern consensus is for a non-reducing, i.e. oxygen-rich, one). Titan is too cold for life, but its atmosphere may contain the complex organic substances similar to those on the early Earth in which the first living entities arose.

To investigate this, the NASA/ESA/ASI (Italian Space Agency) Cassini-Huygens mission to Titan consists of an orbiter, Cassini, and a probe lander, Huygens. After delivering the Huygens probe to Titan, Cassini is now orbiting Saturn during a 4-year period making observations of Saturn's magnetosphere and atmosphere, and also making repeated fly-bys of Titan. These fly-bys and the data acquired by the probe Huygens as it descended through the thick, chemically active atmosphere of Titan revealed its atmospheric composition, which included noble and trace gases, isotopic ratios and complex organic molecules, and imaged the topography of the surface.

When it finally reached the surface, Huygens determined that Titan is covered by a surface of frozen methane and ethane as some

FIGURE 11.13 A composite of a 360-degrees view during the descent of the Huygens spacecraft to the surface of Saturns moon Titan, using 11 of the raw images taken by Huygens. (Courtesy Christian Waldvogel and ESA/NASA/ASI.)

have suggested. It discovered that it has a solid surface with volcanoes and geysers spewing out methane and ammonia or water.

Cassini-Huygens has been a quantum jump in our knowledge of Saturn and its moons, and is revealing chemical secrets of Titan of direct relevance to the questions surrounding the emergence of life in the Solar System.

PLUTO

Although the cold outer reaches of the Solar System are far less likely to support life than the inner planets or the moons of the Great Planets, the exploration of Pluto and the nearly 1000 asteroidal type objects in the so-called Kuiper Belt beyond the orbit of Neptune may nevertheless deliver key clues as to how the planets were born and evolved. To pursue this research, NASA announced in early 2003 a Pluto/Kuiper Belt (PKB) mission for launch in early 2006 and arrival at its targets around 2015. These targets are firstly Pluto and its moon Charon, and then one or even more of the icy bodies of the Kuiper Belt. The half-ton spacecraft will study the surface morphologies of Pluto and Charon, their surface compositions and temperatures, and carry out a detailed study of Pluto's atmosphere. Since there is little sunlight at Pluto due to its distance from the Sun, a radioisotope thermal generator (RTG) is mandatory. This RTG will supply 200 watts of power to the spacecraft and its seven detector instruments.

The objects in this part of the Solar System have never been investigated by any spacecraft from Earth. Since the objects in the Kuiper Belt are believed to be primordial objects dating from the earliest days of the formation of the Solar System (the so-called cold population), and others (the hot population) are believed to have been more recently ejected from the region between Uranus and Neptune, resulting from a slow movement of Neptune away from the Sun, any new information about these objects can be expected to shed new light on the origins of the Solar System and the more recent evolution and dynamics of the outer planets.

COMETS AND ASTEROIDS

Comets and asteroids have different origins. Asteroids come from the asteroid belt between Mars and Jupiter, comets come from the Oort cloud far outside the Solar System. But where did the Oort cloud come from? Huge clouds of gas and dust like the Orion Nebula, just 1500 lightyears from our Solar System, are the cradles of star and planet birth. But this cradle is a violent and inhospitable place. The Hubble Space Telescope reveals the dramatic birth-pangs of young planets in the dust disks which surround newly formed stars. One of the examples of this is the star Theta 1 Orionis C, part of the Trapezium cluster in Orion.

This star produces a huge wind of violent radiation which streams out across space and blasts away at the coalescing dust grains which are in the early stages of new planet formation. It is touch and

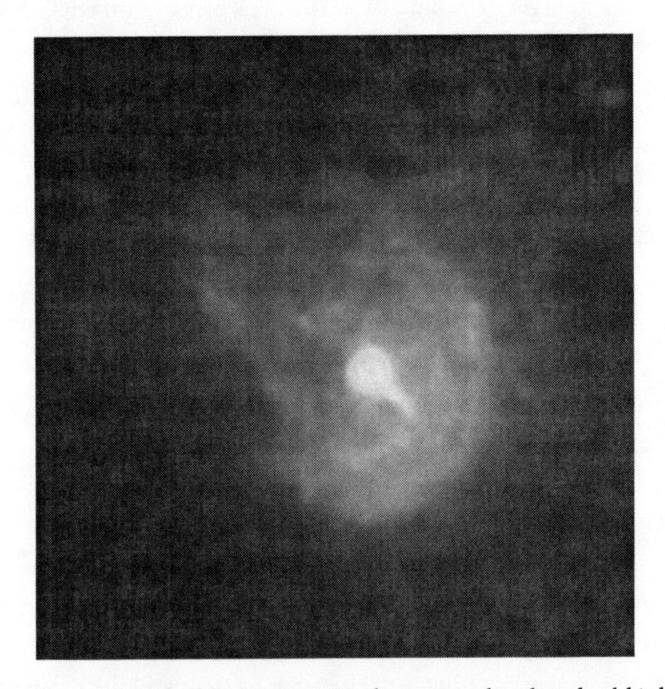

FIGURE 11.14 Theta 1 Orionis C and its surrounding dust cloud, birthplace of new planets and Oort-like clouds. (Courtesy NASA/ESA.)

go as to which process is going to make it. Can gravity win and pull the material together quicker than the blistering tempest of ultraviolet radiation can tear it away? One astronomer describes it as 'like trying to build a skyscraper in the middle of a tornado'.

This violent birth scenario means that only one in 20 of the stars in the vicinity of our Solar System appears to have planets. The blistering flux of radiation from the central star drives huge quantities of matter out into the depths of space far beyond the normal planetary regions, out into regions so cold that they become populated with small rocky objects covered in ice and organics.

These comets appear to be 'dirty snowballs' just as predicted by Whipple in the 1950s. And when they make the long journey back into the inner reaches of the Solar System where our Earth orbits the Sun, they bring with them organics and one of the most interesting substances in the Universe – water. Comets could be the main source of water in the Earth's oceans. Dr Louis Frank believes he has evidence for thousands of small house-sized comets entering the Earth's atmosphere every month. These would deliver huge amounts of water to Earth, thus steadily building up the oceans.

Many of the questions about comets may be answered by ESA's Rosetta mission to the comet 67P/Churyumov-Gerasimenko in 2014, after one Mars gravity assist fly-by and two Earth gravity assists. The mission will consist of a comet orbiter and a comet lander. The orbiter will investigate the large scale composition of the comet, the large scale structure of its nucleus, the flux of dust around it, and the plasma and solar wind surrounding it. The lander will carry a number of analysis instruments for the determination of the composition of the comet's nucleus. It will be the first spacecraft ever to make a soft landing on a comet nucleus. Immediately after touchdown, a harpoon will be fired to anchor the lander to the ground and prevent it escaping from the comet's weak gravity. The analysis of the comet surface and composition will be very similar to what is foreseen for the analysis of the Mars surface. There will be an APX spectrometre, panoramic and microscopic imaging systems, gas analysis and isotopic composition instruments. There will also be a drilling system to drill into the surface to at least 20 cm and retrieve samples. These studies are likely to bring a wealth of data about the composition of comets, their role in the transport of important substances to the Earth and other planets, and potentially their role in the transfer throughout the Solar System of substances vital to life.

If comets brought water and oceans and organics from which Earth life and possibly Mars life evolved, then an intriguing possibility exists. Perhaps comets brought also rudimentary forms of life (like viruses) to the planetary surfaces. Perhaps life itself comes from the cosmos. Decades ago the eminent British astronomer Fred Hoyle suggested that comets might bring viruses to Earth and even be the source of viral plagues such as the global influenza outbreak which ravaged the world after the First World War. The theory was widely rejected at the time, but the more we learn about comets, their links with water and the chemical origins of life, the less and less outrageous these ideas appear.

Slowly a new vision is being born – life as a cosmic rather than a terrestrial phenomenon.

CONCLUSION

Humanity is tentatively putting its toes into the water of planetary exploration. It has not yet committed itself to the recognition that the primary goal of that exploration is the search for life outside the Earth. Short-term planning is in place for the pursuit of robotic missions to Mars, the Great Planets and their moons and some parts of the outer Solar System with probes that have objectives that mix the search for life with other more diffuse goals.

In the missions quoted above and apart from the Human Mars Missions, which in any case have as yet no overall commitment from any space agency, the best estimate indicated is 2015, which is for the arrival of the PKB mission at Pluto and the sample return mission to Mars, hardly more than 10 years from now. Clearly a commitment to human missions to Mars would unleash a huge range of technology developments and innovations and produce fresh thinking, planning and commitments to the search for life on Mars and other planets. In the next chapter we examine what, in an optimistic view, could be possible timewise for humanity's exploration of the Solar System in this century and the search for life.

12 Exploration in time

The exploration of the Solar System in the search for life has also to be seen in a temporal fashion. But first let us see what has been achieved in past and current missions:

MARS

1962	Mars 1 (USSR)	Flew by Mars at 200 000 km – no communications
1964	Mariner 4 (USA)	Flew by – first pictures of Mars
1969	Mariner 6 (USA)	Flew by at ~3000 km – more Mars pictures (no canals)
1969	Mariner 7 (USA)	Flew by at ~3000 km – more Mars pictures (no canals)
1971	Mariner 9 (USA)	First Mars orbiter – sent back 7000 pictures, 100 % of Mars surface
1971	Mars 2 and 3 (USSR)	Imaging, atmospherics, magnetics
1973	Mars 5 (USSR)	Stayed 9 days in Mars orbit – 60 pictures returned
1973	Mars 6 (USSR)	Lander crashed but some atmospheric data sent back
1975	Viking 1 (USA)	Pictures from Mars orbit and first successful lander images from surface
1975	Viking 2 (USA)	As for Viking 1
1988	Phobos 1 and 2 (USSR)	Orbiter/lander
1997	Mars Global Surveyor (USA)	Orbiter, continuously imaging high quality to this day
1997	Mars Path-finder (USA)	Lander and rover produced 16 000 pictures and rock chemical analysis
2001	Mars Odyssey (USA)	Orbiter, imaging, surface composition, water/ice search
2004	Mars Express (Europe)	Orbiter, imaging, surface science, atmospheric science

VENUS

1962	Mariner 2 (USA)	Orbiter, atmosphere/magnetics
1967	Mariner 5 (USA)	Orbiter, space environment
1967–1972	Venera 4–8 (USSR)	Orbiters, atmospherics etc.
1973	Mariner 10 (USA)	Orbiter, atmospherics
1975	Venera 9 and 10 (USSR)	Orbiter and lander
1978	Pioneer 1 and 2 (USA)	Orbiter, remote sensing
1982	Venera 11–14 (USSR)	Landers
1984	Vega (USSR)	Atmospherics, soil sampling
1989	Magellan (USA)	Radar imaging

MERCURY

1973	Mariner 10 (USA)	Fly-by, imaging

JUPITER, SATURN, URANUS AND NEPTUNE

1973–1974	Pioneer 10 and 11 (USA)	Fly-bys, imaging, magnetics
1979–1989	Voyager 1 and 2 (USA)	Fly-bys, imaging, atmospherics
1989–2002	Galileo (USA)	Jupiter orbiter, imaging
2004–2005	Cassini-Huygens (USA and Europe)	Saturn orbiter, imaging; Titan lander, images of Titan surface

This chapter will cover the missions which will take place over the next 20 years, including missions to the polar regions of the Moon and to Mars, the development of key technologies in propulsion, life support systems, and then missions in the next 50 years to the moons of Jupiter and Saturn which will mean a comprehensive exploration of the Solar System in the search for life. Emphasis will be put on predictive methods, how accurate can they be and why is it that we tend to over predict what will happen in the short term and under predict in the longer term, and what this all means for what we can say now about the next 50 years of Solar System exploration.

We will be looking at snapshots in time at the remains of this decade, the next decade, the decades 2020–2030 and 2030–2040 and the double decade 2040–2060.

So, in a nutshell, this looks like the following:

2004–2010

- **MER rovers, Mars Express** at Mars
- Arrival at Saturn of Cassini-Huygens **Saturn orbiter/Titan lander**
- Start of development of **Crew Transport Vehicle (CTV)**
- Start of full-scale development of **nuclear/electric propulsion**
- Launch of **Pluto/Kuiper Belt** (PKB) mission
- NASA **Reconnaisance Orbiter** to Mars in 2005
- Start of development of large **Shuttle-derived chemical launchers** for Moon and Zubrin-type Mars missions
- NASA **Scout** missions to Mars
- Results of **Mars simulation studies**, Antarctic base/bed-rest/radiation/ closed life support systems

2010–2020

- Demonstration of **fully closed life support** systems
- Start of full-scale development of **nuclear/thermal propulsion**
- **EXOMARS/NASA MSL** robotic **search for life** missions to Mars around 2010/11
- Arrival of **Jupiter Icy Moons** (JIMO) **nuclear/electric** mission in Jovian system
- Launch of **Mars Sample Return** mission in 2015
- Restart of **human missions** to the **Moon**
- ESA's **BepiColombo** spacecraft explores **Mercury**
- End of nominal lifetime of **International Space Station** in 2017 extension by 10 years to 2027
- Comprehensive results of **long term survival of humans** in space research on ISS
- Arrival of **Pluto/Kuiper Belt** mission at **Pluto/Charon** system around 2015
- First mars **human tourism** to Low Earth Orbit

- Arrival of **Rosetta** at Comet 67P/Churyumov-Gerasimenko in 2014
- Start of development of **Space Cycler** between **Earth and Mars**

2020–2030

- First **Zubrin**-type mission to **Mars**
- **Subsurface** probe on **Europa**
- Surface **landers** on **Ganymede and Callisto**
- Arrival of **Titan lander**
- Establishment of permanent scientific base at **south pole of Moon**
- Arrival of **nuclear/electric** probes at **Uranus/Neptune**
- End of life of **International Space Station**
- Human **tourism** to the **Moon**
- First **Space Cycler** spacecraft around 2030
- **Comet** and **asteroid** robotic **sample return** missions

2030–2040

- Permanent **human presence on Mars**
- Start of **^{3}He mining** on the **Moon**
- Start of full-scale development of **fusion propulsion** technology
- Start of ***in-situ* resource** utilisation on the **Moon**
- Robotic **mining** missions to **asteroids**
- Implementation of the full **Earth–Mars 'Cycler'** system using three spacecraft

2040–2060

- Start of **terraforming of Mars**
- Start of **biotechnology research** on Europa, Ganymede, Callisto, Titan
- **Human nuclear/thermal** missions to **moons of Jupiter**
- First robotic **fusion-powered** mission to **Mars**
- First use of **Moon** as **assembly** base
- **Human** mission to **Titan**
- **Human** mission to **asteroid/comet**
- Start of **human fusion-powered** missions to the **planets**
- **Mining** of **asteroids**

2004–2010

The immediate period 2004–2010 in the human spaceflight area will be the period of the post International Space Station (ISS) development. Almost all of the industrial development work in the USA, Europe, Japan, Canada and Russia will have been completed. The ISS torch will be passed to the entities charged with operations of the ISS, a work completely different from that of the development of the Space Station. But industrial companies in the USA, Europe and the other advanced countries will agitate for new programmes of hardware development. Politicians will listen to this lobbying. Based on a growing perception of the present Space Shuttle as a flawed vehicle and spurred by the Columbia accident in early 2003, the speeded-up development of an Crew Transport Vehicle (CTV) will be undertaken. This will be dedicated to the transport of humans only to and from Earth orbit. Cargo will henceforward be transported using expendable or partly re-usable vehicles.

Already decided planetary missions will be launched, such as the Mars Reconnaissance Orbiter in 2005 and the Pluto/Kuiper Belt (PKB) mission to Pluto and its environs, which will take place in early 2006. The MER rovers mission and the Mars Express orbiter mission will have been in place since 2004, as will have been the Cassini (orbiter of Saturn) and Huygens (lander on Titan) mission. NASA will carry out a series of scout missions, probably to the Martian poles and possibly a Mars atmosphere sample return mission. The results of studies of simulations of Mars and deep space conditions on humans and robots using Antarctic bases, bed-rest studies, as well as space-based studies looking at the effects of space radiation and the effects of weightlessness over a long period will all be coming on line. The results of technology developments of closed life support systems will be feeding into the design of fully closed systems.

But still the pressure from the industrial complex involved in space infrastructure development will be intense. It will probably first manifest itself in pressure and eventually decisions to engage in the development of the fairly straightforward Shuttle-derived large H_2/O_2

launch vehicles for regular human Moon missions by 2015, and Zubrin-type human Mars missions by the years 2020–2025. The technology will be based on extensions of existing Shuttle main engine technology.

But hot on the heels of all this will be something new, involving a leap in imagination and commitment of the human species to the deep exploration of space. In the timeframe 2006/8, decisions to advance to full-scale development of nuclear electric propulsion systems using nuclear reactors will be made. The nuclear reactors will be used to power Stirling engines that in turn drive electric ion thrusters. These nuclear/electric motors will only have low thrust but will be capable of operating over long periods, making them ideal candidates for robotic exploration missions to the more remote moons and planets of the Solar System. In a sense, there will be a degree of complementarity here, the older chemical system honed to complete its ultimate mission, the human missions to Mars and the newer nuclear systems appearing first in the form of a low thrust long-endurance technology for long flight-time robotic exploration missions. But this will open the possibility of extension to the much higher thrust nuclear/thermal engines capable of the rapid transport of large robotic and human cargoes to and from the planets.

2010–2020

Based on the technological advances and confidence gained with nuclear/electric technology developments and with an increasing awareness of the prospect of planetary exploration, it will become clear that the next step in propulsion technology to facilitate rapid voyages to Mars and beyond will be needed. The start of the development of a technology of nuclear/thermal propulsion, already partially tested in the Nevada desert in the late 1950s and 1960s, will be initiated in this period. Undoubtedly there are many environmental safety and ethical issues intimately tied to this technology. All of these issues can be dealt with in a reasonable and conscientious way. Therefore, the decisions to proceed with the full-scale development of this technology will be taken *relatively* easily. Only the more extreme elements of the anti-nuclear and environmentalist movements are likely to protest.

Similar technological advances are expected in the crucial area of life support systems. Completely closed systems that greatly reduce the demands of a wide range of consumables and probably based on bioregenerative processes will be demonstrated in this period.

In this period the robotic missions to Mars will multiply. The ESA EXOMARS mission, with a specifically designed 'search for life' rover drilling and instrument package, will be active on Mars. NASA plans a similar mission at that time with its Mars Science Laboratory. A sample return mission or missions will take place around 2014/15, demonstrating Mars orbit rendezvous and safe return to Earth.

The detailed analysis of the samples brought back may greatly deepen our knowledge of the Martian surface/subsurface and answer questions about possible life on Mars. Around the same time ESA's BepiColombo will be scouting out Mercury – the first such probe to do so since Mariner 10 in 1973.

The investments in the development of nuclear/electric propulsion technology already engaged in 2006 will result in the launch of the first nuclear/electric robotic mission to Jupiter's icy moons that will arrive in the Jupiter system in about 2014/15.

By this time we can expect a restart of human missions to the Moon. This will be driven by a realisation that the Moon is not an end-point, a target, an objective to be reached and then abandoned in the Apollo 1970s way, but rather a very useful way-house, source of resources and assembly point for further penetration into the Solar System.

After 10 years of operations, the International Space Station (ISS) will be at the nominal end of its life. Certainly the contribution of the ISS to research in physical and life sciences, as well as use of the ISS for many astronomy observation objectives, will have proven useful. But nonetheless, nominal termination of lifetime will be declared. It seems likely, however, that in view of probably interesting results on the survival capabilities of humans in long-term space conditions and the prospect of future more interesting results, the decision will be taken to prolong the life of the Space Station. This will be similar to the case of the MIR space station in the 1990s when Russia, despite the collapse of the Soviet Union which had provided such extensive resources to support large-scale space activities, managed to keep the MIR space station in orbit until the late 1990s. This is in contrast to the USA experience with its post Apollo, Skylab space station that lasted manned only about a year or so before waning public interest and an unfavourable political climate ensured its demise, despite many interesting scientific results from its flight(s). Optimistically then, we can expect an extension of ISS on-orbit operations of about 10 years to the year 2027.

Marking symbolically the outer reaches of human exploration of the Solar System, the year 2015 will see the arrival at the Pluto/Charon system of the Pluto/Kuiper Belt (PKB) mission. This will be a purely chemical propulsion mission with an RTG generator of electrical power supplying about 200 W. The surface morphology, atmosphere and other features of Pluto and its moon Charon will be mapped.

The commercial exploitation of space will take off in this epoch. An increasing number of wealthy/well-off individuals will have at first enjoyed million dollar sojourns to low Earth orbit (LEO). However, as time progresses, flight costs will decline, certain individuals will get more well off by the natural processes of capitalism, and we will arrive at a situation where the costs of space tourism, especially to low Earth orbit, will hardly be prohibitive for a substantial number of wealthy persons.

By this time, sophisticated methods of calculating celestial mechanical gravity assist trajectories, of designing high performance closed life support systems and of flying reasonably priced taxi flights to orbital systems will allow the start of development of Earth–Mars 'Cycler' systems. These will be autonomous spacecraft permanently in orbit around the Sun but passing by on a regular basis both Earth and Mars. This will allow access by taxi flights to and from these planets to the Cycler, which provides the main transport between the Earth and Mars and back (see Chapter 14).

2020–2030

By 2020–2025, the die of future human exploration of the Solar System should be cast. We can expect the first Zubrin-type human Mars missions around 2025. The mission and follow-on missions will be based on the proven capability to create out of the CO_2 atmosphere of Mars the methane CH_4 and oxygen O_2 to serve as propellant and oxidiser as the return fuel to Earth. The consequences of this are that semi-permanent human occupations of Mars will impose themselves in the people's minds. Planning will start for the semi- and, ultimately, completely autonomous operation of this type of base.

Robotic missions carried out in parallel will continue apace. By 2020, subsurface probes from robotic missions should be operating on the icy surface of Europa and possibly Ganymede and Callisto. These probes will be boring into the ice of these moons looking for the point where the liquid water or the slush starts and then making microscopic analyses of the water/slush to see if any microbes or microscopic life forms are present. One can imagine the significance of the discovery of such life forms. The Earth/Mars system is clearly a closely interacting one and the non-discovery of Earth-like microbes on Mars would be a surprise, given what has been discussed in earlier chapters about the interchange of matter between the two planets (see Chapter 5). Yet the organisms, although they may have originated from Earth, might have evolved in a completely different way from any Earth organism and as such would be of supreme interest from a genetic point of view. But Europa is almost certainly very weakly coupled with either Mars or Earth. So the discovery of microbes beneath the icy surface of Europa would be a significant event.

The Cassini/Huygens mission to the Saturn/Saturn-moons system will give many clues as to the atmospheric and surface properties of Titan. It has discovered that beneath the thick nitrogen and methane soup of the Titanic atmosphere there are landmasses and oceans. In which case, there will be an immediate interest in re-visiting the Titan surface with another lander. The results of the Cassini/Huygens mission in the year 2005 will have pushed the scientific community to promote another Titan mission, this time more ambitious, more exploratory. Around 2020, a Titan lander/explorer will soft land on this moon and start the search for micro-organisms.

Also during this period, a permanent base at the south pole of the Moon will be functioning. Crews at the base will start to make experiments on the extraction of oxygen and possibly minerals from the lunar soils, as explained in Chapter 10, and forays to any local ice formations to investigate if this ice can be used as a reliable source of water for such bases. It is also possible that the first experimental extractions of ^{3}He will start about this time.

We can expect a comet sample return mission in this decade based on what is learned from ESA's Rosetta mission. This should allow a definitive assessment of the organic and possibly biological content of the surface of comets. Asteroid return missions are also likely in this period.

The motivation and desire for further exploration will probably be strongly enhanced by the discoveries made to date on Mars and on the moon systems of Jupiter and Saturn. Nuclear-electric propulsion technology will by this time have proven itself and the combination of these factors will lead humans to go further afield in search of new discoveries. To this end, the decade will see nuclear probes arriving at the moons of Uranus and Neptune.

2030–2040

The Zubrin scenario envisages an immediate permanent presence on Mars from the first landing onwards. However, it seems unlikely that that scenario will attract the approval of Earth-based planners. More likely is a situation in which there is an intermittent presence of humans on Mars over a period of years. Since it will be as onerous to transport people back to Earth as it is to bring them to Mars there will come a push for a permanent presence on Mars. But then immediately will arise complicated logistical and ethical issues about how long people stay on the planet.

Like other exploration scenarios played out on Earth, there will come a psychological moment when it is realised that there are people/explorers *who will not come back to the Earth.* This will be totally contrary to the thinking inherent in President Kennedy's speech of 1962 when he vowed that America would 'send a man to the Moon *and bring him back safely to the Earth'.* This means that at some point some human beings will go to Mars and stay there without any prospect of returning to Earth in their lifetime. Indeed, there will arise the situation where people are born as 'colonists' or 'planetaries' who never have the possibility or, in extreme circumstances, the right, to return to the 'home' planet.

It is difficult to downplay how significant culturally, psychologically and emotionally this moment will be in the history of our species. Coupled with this momentous decision will be another one, the decision to consider the base at which one is living, not as a base from which one could eventually escape, but as a base that is home. This will represent a major shift and transformation in the way of fundamental thinking of a small but highly significant number of members of our species.

The permanent presence of humans on Mars will depend on the taking of that decision. It may well be that this will attract and/or involve people from philosophically committed cults and sects prepared to make deep sacrifices for a new life. One immediately thinks of the *Mayflower* that arrived on American shores in 1620, bringing a group of people fleeing from religious persecution in Europe, practically destitute and with no real hope of return. One can imagine a similar situation in the middle of the twenty-first century. A committed group of people make a pact with a commercial company for a Zubrin-type series of missions to Mars, clear in the intention to populate and possibly terraform Mars based on a particular religious or apocalyptic world view. At that point, the bureaucratic/technocratic hold on the future of space exploration will be broken, never to be re-established. Space exploration will pass into the realm of what can be achieved by ordinary people working in groups, outside the control of governments either at national, regional or global levels.

In the meantime, the feasibility of ^{3}He extraction from the Moon is proven and the helium mining is initiated. The helium is liquefied at the Moon base and stored in tanks that are brought on a regular basis back to Earth, where the helium is burned in ^{3}He–^{3}He third generation fusion reactors to generate a significant portion of the Earth's energy.

In the light of the Moon-Earth helium success story, experimental designs are evolved for fusion-based rocket propulsion systems. These will be even more powerful than the nuclear-thermal systems and offer the means for relatively straightforward traffic between the bodies of the Solar System. The vehicles will probably be fuelled-up in low Earth orbit with helium coming directly from the Moon.

By this time, human tourism to the Moon will have become a multi-billion dollar business with privately owned and operated hotels on its surface.

The decade will end with the placing of a complex of three spacious and sophisticated spacecraft in special orbits around the Sun as the Earth-Mars Cycler system. Taxi flights using chemical rockets will transport people and cargo to and from Earth and Mars to the Cycler spacecraft.

2040–2060

It's the year 2060, nearly a century after the first men set foot on the Earth's Moon, the first planetary body to be touched by human feet. On the launch pad sits the huge hydrogen/oxygen rocket that will bring the nuclear/thermal assembly into low Earth orbit. There are three astronauts on board and these three will be the crew of the vehicle that will bring them to Saturn's moon, Titan. With its heavy atmosphere of methane and nitrogen and the discovery in 2043 of the fact that the Titan ocean has certain microbes in it, Titan has suddenly become a point of interest for the Earth-bound biotechnology community. By sequencing the DNA of the microbes, it had become clear that a range of interesting bacteria from a biotechnology point of view might exist on Titan. The mission is to search for these micro-organisms that could be of highest interest to the burgeoning biotechnology industries of the twenty-first century.

The countdown continues. The huge Shuttle-like vehicle finally vents its flames onto the Florida soil and climbs slowly and noisily into the sky. Seven minutes later the chemical engines shut off at an altitude of 250 km. The crew adapt to the microgravity environment. Twenty-one hours later a nuclear reaction is activated. A steady stream of hydrogen is forced through the guts of the reactor and a powerful surge is generated, swiftly pushing the human spacecraft out towards the moons of Saturn. Titan is only months away. The crew prepare themselves for the long voyage to Titan, for the relatively short stay on the Titan coastline and the difficult biological analyses that they will have to do to select the samples they wish to bring back.

Fanciful this may be, but the discovery of unique forms of life on another world will open the prospect of some parts of the Solar System harbouring micro-organisms that might be of interest to the large biotechnology industry now emerging. The PCR reaction widely used today to amplify DNA was developed from extremophiles, and pharmaceutical companies might in future be prepared to support missions to biologically interesting parts of the Solar System to search out new organisms.

The period 2040–2060 will see other significant developments. The first limited steps in the direction of terraforming may be taken. Experiments to grow plants in Martian soils and to test various bacteria for their ability to create atmospheric gases from Martian soil will be carried out, probably in greenhouse environments.

By about 2050, human nuclear/thermal mission capabilities should have been established and proven by human missions to the moons of Jupiter and, by the end of that decade, as seen above, by the human mission to Titan. A proof of concept of a fusion propulsion system will be tested on a robotic mission to Mars and, by 2060, human fusion powered missions should be starting to become possible. The development of these important technologies will start to render economically viable activities such as asteroid mining, and the use of the Moon as an assembly base.

CONCLUSION

Fantastical as some of these ideas might seem, given the capabilities of today's technology, none of this should be particularly difficult to achieve. It should also be kept in mind that the history of the predictive arts usually shows a tendency to overestimate progress in the short term and underestimate it in the long term. So some of the things we have predicted here for 2060, distant in prospect as they may appear to us now, may well be the commonplace of spacefaring humanity in the middle of this century.

13 Prediction, imagination and the role of technology

Finally, our coming ability to maneuver in interplanetary space will rival the exploration of the New World in producing an age of discovery and excitement. Within fifty years we will probably have bases on the Moon, on Mars, and circling Jupiter. These bases will act much like the first European outposts in the New World in the fifteenth and sixteenth centuries, producing a steady stream of wonders that stimulate our imagination and curiosity. And being 'out there' greatly increases our chances of receiving evidence (à la SETI) of other civilizations in our galaxy. Such an observation, if made, will have effects at least as great as the effects on medieval Europe of the rediscovery of the works of classical Greece.

John H. Holland (2002)

No perspective vis-à-vis the future can avoid dealing with the three elements of prediction, imagination and the role played by technology. As we have said, prediction about the future in a technological or scientific context can in a classical way be somewhat similar to the predictions of economists and stock market analysts in financial and economic matters. This classical approach is essentially passive in nature, projecting future trends based on past performances and modulated by the often very subjective biases of the predictors. John Holland, a Santa Fe Institute member and professor of psychology and computer science, and his work represent maybe one of the best examples of new ideas in predictive theory. His ideas are focused on what are called Complex Adaptive Systems (CAS), consisting of many interacting components called agents that adapt to (or learn from) each other as they interact.

Holland focuses in one of his essays on the extraordinary difference between advances in computer hardware capabilities where these capabilities double every 18 months (known as Moore's Law), as compared to the vastly slower doubling time of between 10 and 20 years for the proficiency of software. This is a stunning difference. It means that over a period of about 20 years, say from 1975 to 1995, computer memory and processing capability increased by a factor of 2^{13} or about a factor of ten thousand billion, whereas software capability only increased by between 2 and 4. Truly the triumph of brawn over brain.

A similar theme is taken up by computer science guru Janon Lanier, who also identifies software as the Achilles' heel of the computer revolution: 'Accompanying the quixotic overstatement of theoretical computer power has been a humiliating and unending sequence of disappointments in the performance of real information systems. Computers are the only industrial products that are expected to fail frequently and unpredictably during normal operations... Specifically it is the software that seems impossible to manage for a predictable price.' But who could have predicted all this in 1970, for example? Probably nobody. But today, as a result of the work of John Holland and others, Complex Adaptive Systems theory is helping to hone the accuracy of predictive methods. As can be seen from the quotation at the start of this chapter, the predictions of Holland based on these methods is consistent with the ideas we have presented in the previous chapter regarding the prospect for future exploration of the Solar System by the human race.

When we turn to the role of imagination we see that all great feats of exploration have been based on the human ability to imagine new worlds and the challenges and opportunities they offer. Combined with the will to devise and push through the means to carry out such exploration, this type of adventure adopts a mantle of near unstoppability. As Hannibal said, 'We will either find a way or make a way,' and finding a way or making a way will be the story of future space exploration. But will and imagination are usually the province of individuals or small groups of men motivated by powerful visionary, mercenary or even religious ideals.

In Renaissance times and the ensuing centuries, the voyages of exploration were carried out by private entrepreneurs, usually with a combination of private and local prince/king support. Probably the same scenario will apply to the successful voyages of discovery by the Solar System exploration pioneers of the twenty-first century.

Certainly the world's space agencies, such as NASA, ESA, NASDA and the Russian Space Agency, have shown little inclination to promote and support voyages of exploration by private corporations. This will change in the early twenty-first century. New technologies are emerging to reduce the cost of launching and sending robotic and human missions to explore other planets of the Solar System and set the stage for exploitation of these planets.

But if we accept that history will repeat itself, that the objective facts of exploration, exploitation and colonisation remain unchanged, whether it is the sixteenth century and the discovery and colonisation of the New World and other territories, or the twenty-first century and the exploration and colonisation of the Solar System, we conclude that the latter will probably be best carried out with a mixture of private corporate and public funding.

Let's look at some examples of exploration enterprises and see if we can identify key aspects that can serve as common points with future space explorations:

- Magellan's first circumnavigation of the Earth in 1620.
  Magellan was a visionary, a zealot and a man supported by a king, the King of Spain. His vision drove him to find a way through the straits named after him and into the Pacific Ocean and ultimately the first circumnavigation of the globe. His support from the king got him the means to raise enough capital to outfit three ships as the minimum needed for the voyage (only one returned to Spain), and his zealotry got him killed in the Philippines while trying to convert the natives to Christianity.
- Captain Cook's voyages of discovery in the eighteenth century.
  Cook was an inquisitive and brilliant navigator backed by the British Admiralty. He was no visionary but was a practical and persistent explorer prepared to take calculated risks with the backing of the British government.
- The US lunar Apollo missions, 1969–1973.
  This programme was a classical example of a big government, big spending, programme with the private sector involved only as

contractors to government (NASA). Therefore, all the risks were taken by government, the commercial risk exposure of the contractors being limited by essentially cost reimbursement contracts. And the sole reason for all this was that the programme was a national prestige programme – even with aspects of national security involved, since it was carried out at the height of the Cold War with a strategic purpose in that war.

So, what are the lessons to be learned from these examples? They seem to boil down to:

- Governments will not accept all the risks for these types of activities unless issues of national security or national prestige are involved.
- Private corporations also cannot cover all the risks of these new enterprises.
- A combination of risk sharing between private corporations and governments seems to have the best chance of working.
- Visionaries, persistent and 'unreasonable people' are needed to push the enterprises through. These individuals should play leadership roles in promoting the enterprise or in carrying it out, or both.

Let's now turn to technology. Today we have become used to the notion that a very significant portion of the global economy is based on or related to technology. In a sense, we expect that technology will, bit by bit, take over our normal working environment and economic activity. In the case of space research and exploitation, technology is king. The question then arises if each technological development has a pace of its own determined by the technology in question. If so, it becomes a major constraint. Put simply, it means that we cannot do anything until the technology 'becomes available'. Reflection leads us to the conclusion that technological development is driven by a number of factors, some or all working at once:

- Human inventiveness/ingenuity usually linked to a perceived need within a commercial market.
- Human crises (war/revolution, natural catastrophe, survival issues).

- Human acquisitiveness/power seeking, desire to profit/dominate.
- Human desire for security (technologies developed within defence budgets).

Further reflection on these points will reveal that in almost all cases the rate of technological change or progress will depend on the strength of the key factors at play, the pressure to come up with the required technology and the resources, both human and financial, devoted to the problem as a result of those pressures. So it seems technological progress is determined by needs and by the resources that are made available. A striking example was the Manhattan Project, where, after the US government had allocated all the resources possible in time of war, it led to the development of the nuclear technology to construct nuclear bombs in a period of 3–4 years compared to a likely development time in peacetime of perhaps 10 to 15 years.

So how can we predict what will happen in any given context? Clearly, the Complex Adaptive Systems approach of Holland and others described above will be relevant here, and this type of analysis is working its way into present-day technology predictions.

One of the major aspects of predicting technology-driven change is the observation that, as stated above, we tend to *overestimate in the short term* and *underestimate in the long term.*

There happens to be a very simple explanation of this phenomenon that is related to the difference between linear growth and exponential growth. Since technology growth as we have seen above is driven by needs and resources, it is in this sense an economic activity. By and large, humans tend to think in terms of linear growth, which is based on the idea that growth is a step by step linear process. But actually, all growth in the biological and economic worlds is exponential in nature in the first part of the growth cycle. In the latter part of the growth cycle it tends towards some asymptotic limit, either external or internal. So the overall shape is an S-curve or sigmoid representing the difference between the linear and

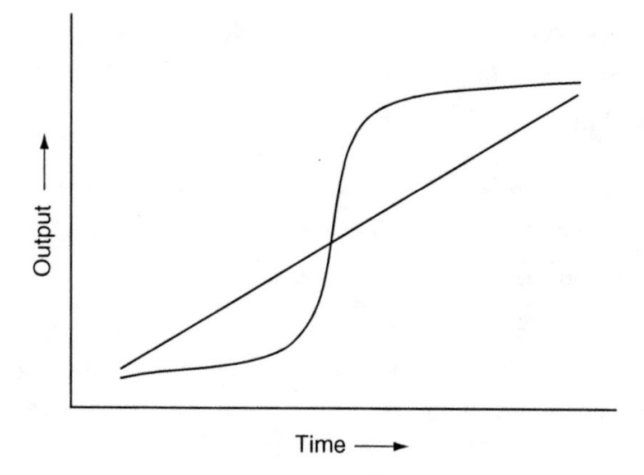

FIGURE 13.1 Sigmoid (first stage exponential, latter stage asymptotic) and linear growth curves.

exponential models that we see in Figure 13.1. In the near term, the linear prediction will outstrip the early part of the sigmoid. In the mid term, they are roughly in line, but in the longer term the sigmoid outstrips the linear projection.

So in terms of what technology offers in the pursuit of goals, we can see it as:

- Enabling – when in the full growth part of the S-curve
- Disabling – when nearing the limit part of the S-curve

Now the right part of the diagram represents the approach to a technological limit for both the sigmoid and the linear growth curves. Examples of technological limits are:

- Speed of sound (thought to be unbreakable by aircraft until proved otherwise in 1947)
- End stage of Moore's Law (given by physical limits of miniaturisation on silicon and other semiconductors)
- Solar panel efficiency of 15–20 % (determined by inherent characteristics of semiconductor materials used)
- Chemical propulsion (limited by chemical properties of propellants used)

All of the above can be transcended by switching to, or developing, new technologies. This is unlike physical limits, which are absolute. Examples of physical limits are:

- Speed of light (absolute limit of velocity for all objects in the Universe)
- Second Law of Thermodynamics (no energy creation *ex-nihilo*, no perpetual motion machines, limits to efficiencies of all processes at given temperatures, increase in entropy)
- Quantum mechanics (Uncertainty Principle, limits to accuracy of simultaneous measurements of energy and time or of momentum and position)

So the conclusion we are heading towards is that to do certain things utilising the laws of physics or biology to our advantage then technology is decisive. If we don't have it we can't do it (as it were). But, and it's a big 'but', if what we want to do transcends only existing technological limits and not any hard physical, biological or computational limits then we can, and humanity probably always will, if the needs are strong enough, allocate the resources to develop the technology to accomplish any task or goal it sets itself. The two main resources needed to propel technological development are:

- Time
- Money

In general, history shows that in peacetime it is time which determines technological progress – the time for inventors and market demand to produce the inventions which drive technology development. In war or national emergency as we have seen in the example quoted above, it can be different. We can have a matrix:

	Time	Money
Peacetime	Plentiful	Scarce
War/National Emergency	Scarce	Plentiful

As examples, the Second World War saw unparalleled financial resources deployed by the USA and other governments to drive technological advances for the war effort (jet engines, rockets, radar, computers, atomic bombs) in a situation where time was of the essence, i.e. scarce.

Let's finish off this section by talking about money. Comparisons are often made between what is spent on defence on the one hand and more peaceful research and development activities on the other *on an annual basis*. Yet it is probably more revealing to compare these numbers on longer timescales, such as over a few decades. The United States spends about $400 billion per annum on defence. This amount is more than the rest of the world combined, which is estimated at an additional $300 billion per annum. This gives a total global defence spending of circa $700 billion per annum. So over a decade this is $7 trillion, over 30 years $21 trillion, and over a century $70 trillion.

These are truly staggering amounts compared to what at first sight appears to be the excessive costs of space endeavours such as the International Space Station. This project is the largest cooperative international scientific, technological and engineering project ever undertaken. It was started around 1985 and is destined for 'Assembly Complete' around 2006/7. With a nominal operational lifetime of about 10 years, a total development and operational cost of $100 billion over 30 years has been estimated.

This can be compared to the global spending on defence of $21 trillion over 30 years, i.e. it is about half of one per cent of that spent on defence. Estimates by Zubrin and by NASA studies for a large-scale series of exploratory and early colonisation missions to Mars also stand at about $100 billion over 20 or 30 years. So for just 1 % of the resources devoted to global defence over the next 30 years or so, a large International Space Station *and* a vigorous Mars exploration and start-up colonisation programme would be possible. Adding the same amount again, about 2 % of the total global defence spending over 30 years would probably deliver technological breakthroughs in nuclear propulsion for fast interplanetary travel, new developments in electric propulsion, sophisticated missions to the moons of Jupiter, Saturn,

Uranus and Neptune, advanced life support systems for planetary colonisation, and low Earth orbit space stations for assembling interplanetary mission infrastructures.

And these are only the technological products. Fundamental research in physics, biology and medicine on platforms such as ISS are already producing results applicable to creating new materials and processes, as well biomedical results applicable to health care on the ground. We have already seen the likely high interest of pharmaceutical and biotechnology companies if unique organisms are found in other parts of the Solar System.

It is often stated that politicians can only decide on issues which will be hot at the next or possibly the next but one election, i.e. over a timescale of only 3 to 5 years, and hence long-term funding decisions needed for the type of enterprises argued for here are beyond their horizons. Yet a cursory survey of political procurement decisions in the defence sphere shows that it is increasingly common for procurement decisions to be made that involve tens of billions of dollars or euros covering timescales of over 20 years. So, to exaggerate a little, the decision to develop and deploy, for example, a new range of aircraft carriers and the attendant battle groups can take 3 to 5 years to be arrived at, have a development, construction and commissioning time of 10 to 15 years, and an operational lifetime of 15 to 20 years, giving a lifetime total of between 28 and 40 years. So the arguments hinted at above for long-term decisions on space exploration spanning 30 year timescales are perhaps not so wildly divergent from emerging practices in the defence sphere.

CONCLUSION

Reliable predictions regarding future developments in space exploration and technology are not easy to make, but there are general features regarding motivations and policies that can be discerned similar to those seen in exploration periods in the past. Sophisticated analysis techniques as described by Holland can be used for prediction in this area and appear to produce results broadly in line with the predictions made in this book. In general, there is a tendency to overestimate in the short term and underestimate in the long term. Technology can be a stumbling block if you don't have it, but once you do, it is rarely a limiting factor. If the need is sufficiently powerful, time and money will usually allow ways to be found to transcend 'limits' of a technological nature. Much of these technological investments can profit from combined and complementary stakeholdings by private corporations and by public funds. Finally, in a decision-making world dominated by technology investment, politicians already face procurement decisions in the defence sector with timescale impacts similar to those that will be required for the initiation of the bold space exploration programmes dealt with in this book.

Part V
Our cosmic destiny

What, after all, is the alternative? We can either stay at home, sending a few robot spacecraft to our neighbouring planets, and continuing to gaze at the more distant universe across light years of empty space, or we can get ourselves out among the planets and, eventually, the stars? In which alternative future would we learn the most about this universe and our place within it?

I. A. Crawford (1998)

Surely if humanity can spend trillions on drugs, armaments and advertising, it can marshall a few billion for a deeper, more complete understanding of its place in the cosmos.

Jean-Michel Cousteau (1999)

14 Our cosmic destiny

The question of whether or not there are or ever have been other viable life forms in our Solar System and whether or not there is any prospect of finding other intelligent life in our Galaxy will, with a high degree of probability, be settled by the end of this century. *This is a stunning prospect.* It means that by the time the grandchildren of people aged 30 today are laid to mortal rest we will know if we are alone in our Galaxy or not.

At that point, the path taken by *Homo sapiens sapiens* 40 000 years ago when he stopped being a hunter-gatherer like his primate ancestors, and became a non-nomadic farmer, an investor in place, in buildings and in civilisations, will reach a crossroads. Humans, up to now content to see their world bounded by Earthly horizons, who live in cities as civilised beings along with their fellows under the rule of law, but who nevertheless build telescopes and high energy particle accelerators to seek the truth about the Universe, will have arrived at a fork in the road of their destiny. At that point our species will be faced with fundamental choices.

The most significant of them will be moral choices. If there are no other life forms found in the Solar System, do we have the right to populate it with ourselves and potentially use Earth micro-organisms to terraform other planets? If we do find life in whatever form on other planets of the Solar System, what rights do we have to use it for our purposes and what obligations are on us to protect it and its habitat? If we find no traces of other intelligent lifeforms in our Galaxy, does that give us the right to lay down plans for the aggressive invasion of the Galaxy by our species, a colonisation which would take by all reasonable calculations a mere 100 million years, or just 1 % of the age of the Universe? Or perhaps this galactic prospect is just too abstract to stimulate any reasonable debate about how we should act.

In any event, the starkest choices will be the immediate ones about our exploration of the Solar System. In a sense we are already entangled in these choices. The guidelines we are laying down and following now regarding planetary protection and covering the robotic missions that are already under way are the first steps in facing those choices.

No one today seriously questions the future of humankind in the exploration of our Solar System, the Galaxy and the broader Universe, including the deepest laws of physics, chemistry and biology that inform our existence on this planet. But the real nitty gritty question is what role the humans themselves should have in the exploration? Strong opinions have existed for nearly half a century on both sides of this debate. Broadly speaking, scientific commentators in the 'hard' observational sciences, such as astronomy, cosmology and high energy physics, have tended to be against human spaceflight as a means of discovering new science. On the other hand, a broad range of 'soft' science and philosophy people, such as biologists, system theorists, geographers, and philosophy and history of science people, have been in favour of space exploration by humans. As an exception, and notably, 'hard' science geologists have been consistently *for* it, as they often quote the mantra of the importance of the 'geologists in the field'.

Turning first to those 'hard' scientists who were and continue to be against, the first definitive round in this battle was fired by the British Astronomer Royal, Richard Woolley, in January 1956, when he dismissed space exploration and particularly the possibility of interplanetary travel by humans as 'utter bilge'. During the build-up to the Apollo landings on the Moon the argument festered on, many in the 'hard' scientific community pronouncing highly dismissive judgments on the scientific basis of what they perceived as the purely political motivations for the Apollo programme. Woolley again proclaimed just before the Apollo 11 first Moon landing that 'from the point of view of astronomical discovery it is not only bilge but a waste of money'.

British academic Ian Crawford has eloquently demonstrated that the scientific payback from Apollo was in fact quite significant despite these claims. He arrives at an important conclusion, i.e. 'Any space mission that has to transfer people will, by its very nature, be able to carry a significant scientific payload, even if science is not the primary driver for the mission'. This insight is fundamental and expresses a key notion. Substantive investments in exploration and ultimately colonisation will never be made for purely scientific reasons. The reason for these considerable investments will be more likely found in humanistic, political, economic, commercial and cultural motivations and impulses, reflecting science as a significant but not overarching human preoccupation. Crawford's insight about this historical aspect of spaceflight he extends to make a general point: 'It seems to me that most of this opposition ... stems from two implicit but erroneous, assumptions: first that the primary motives for sending people into space are, or at least ought to be, scientific, and second, that the high cost of human spaceflight is taken from existing scientific budgets.' Crawford, in our view, effectively dismisses these assumptions and builds a powerful case to argue that science has a lot to gain from hitch-hiking itself to human exploration missions decided, in any case, for other more pressing political and economic reasons.

An entirely different rationale comes from American engineer Darrick A. Dean. Focusing on the idea of a frontier, a horizon and a future target for the exploratory instincts of humanity, he argues that without such horizons human civilisations fall into apathy and stagnation. Dean's arguments are hardly based on any substantive historical analysis of the decline and fall of the Roman, Babylonian and Egyptian empires, which he quotes as evidence for his thesis. But interestingly, he itemises some aspects of our present sociopolitical conditions which he sees as evidence of decline, such as the increasing bureaucracy in government, education and business; the inability of government or industry to undertake great new projects (cancellation of the super-collider, inability to solve or radically decrease social problems despite immense wealth, continuous delays in advanced transportation systems); the spread of irrationalism and pesudoscience (tabloid science, astrology, alien abductions, government conspiracies); and the decline in the rate of significant innovation (as compared to past decades).

These ideas are perhaps somewhat simplified versions of reality. For example, the degree of innovation and risk taking in the IT industries in recent decades has been unprecedented.

Another aspect of this discussion is the *supposed* circularity of the arguments in favour of human spaceflight. Typical of this kind of argument is the statement that it is ridiculous to do research on human long-term survival in space since this is based on the assumption that humans are being sent into space in the first place. The circularity exists only in the minds of those who oppose human spaceflight. Basically they argue that since there is no clear case for humans to venture any further into space or the planets, there is no need to do long-term research and studies on the effects of the space environment on humans. But this is also a circular argument, and it ignores the fact that research per se on long-term exposure to space has significant results for ground based health issues, such as osteoporosis and muscle degeneration in the ageing.

What seems to be always in dispute is the desirability of human spaceflight, which is less so criticised regarding low Earth orbit (LEO) but still continually sniped at by various elements of the scientific, literary and artistic world but not usually, and interestingly, the media, political or technology worlds. It seems almost as if there is a 'two cultures' effect similar to C. P. Snow's argument in the 1950/60s between the literary/artistic and the scientific worlds. But curiously the two sides of this argument are polarised somewhat differently. The split here is the fundamental science/artistic/literary camp versus the geopolitical/media/technology axis. Interestingly, the science fiction element is probably a deserter from the first camp to the second, and probably the military from the second camp to the first.

Added to this argument is another, more recent one, which focuses on the role of space agencies such as NASA and ESA, seeing these bureaucratic government owned institutions as impediments to space development and commercialisation. This set of arguments has been characterised as 'libertarian' and chiefly sets out to prove that space agencies should concentrate only on basic research and technology development, and contract out to industrial/commercial companies the final development and implementation tasks. This is quite strongly argued by American economic commentator Lou Dobbs in his book *Space, the New Business Frontier*. Along these lines, American commentator John McKnight, basing his arguments on the ideas of Robert Nozick, professor of philosophy at Harvard, argues that organisations like NASA are constitutionally incapable of promoting advanced human exploration and exploitation of space. McKnight bases his argument on his categorisation of the fundamental split between the agendas of the US 'space movement' and NASA as utopian versus anti-utopian forces. Placing Plato, and more recently Zubrin and G. K. O'Neill, who advocated space colonies in the 1970s, in the US space movement utopian camp, he identifies NASA as a classical anti-utopian organisation. He argues that utopians believe that society is capable of being improved and that the means of this improvement will usually involve deep structural and institutional changes within that society, including the shift of power and influence away from some groups in favour of others.

Since NASA is a branch of the US government and hence a tool of the realpolitik of that government it must be a mechanism for promoting the political agenda of that government, which clearly has a stake in the status quo and certainly no interest in promoting radical structural changes in society. This is not to say that people working for organisations like NASA have no utopian ideals – some may even work on utopian projects on the margins in those organisations – yet the organisation as a whole cannot act against its primary purpose. McKnight cites as examples of this the abandonment of the Apollo missions and any further Moon activities in the 1970s, after they had served their purposes in facing down the Soviets in space, and the vehement opposition of NASA to the flight of Dennis Tito on the Space Station. Tito had clearly stated his utopian ideal, namely his ambition to fly as 'a product of individual means for individual motives, in direct opposition to the ISS's purpose as a tool of collective state diplomacy' The Russians, of course, who had the means to fly him on their Soyuz spacecraft, were only too willing to oblige these ambitions. Interestingly, international organisations like ESA may be less so constrained, especially if mandated by the sometimes utopian ideals of the European Commission. As far as NASA is concerned, the debate following the Columbia tragedy has focused attention on the role of NASA in US space activities and may result in structural changes, allowing the more entrepreneurial instincts in US society to find a role in the future of human spaceflight.

The existentialist maxim says that we are condemned to choose. Conceived as a statement about the predicaments of individuals faced with a complex world of moral, ethical, sometimes economic and even technological issues, it now presents itself as a predicament to us as a species. We are used to thinking of choice as an individual matter, often as a matter of conscience that we perceive almost always as a personal rather than a group issue. Yet there are choices that are not choices of conscience. Choices about the survival of our species. Choices about the best way forward for our civilisation and species. Choices that we will have to make at a species level. And, just as for an individual, we can never choose, anticipate or plan for the choices that we are ultimately faced with. Often the most excrutiating choices are concerned with newly opening opportunities rather than newly diminishing ones. And that is where humanity finds itself today. For the first time in the history of humankind, by virtue of all that has happened over the last few centuries and above all the twentieth century, we are, as a species, faced with such a choice.

And the choice is this: do we want to pursue our future in space through the means of status quo politics, calculating every move and investment on the basis of risk-aversion, cost benefit analyses, increasingly driven by introspective instincts, managed by realpolitik-driven organisations, our opinions formed by the spokespersons of an embedded cultural elite antipathetic to technological and social progress and risking the probable fate of our civilisation withering from within? Or are we prepared to see the search for life as a truly historic challenge whose time has come and which beckons us to make the brave decisions that will lead to the spreading of humanity's footprints into the broader reaches of our Solar System?

The search for life beyond the Earth is a noble cause. The question of whether or not life on Earth is a unique local phenomenon or life can be found elsewhere in our Solar System and Galaxy is one of the most fundamental scientific, philosophical and existential questions of our time.

If we can identify a way to answer this question, we are almost morally obliged, as a duty to future and even to past generations, to pursue that way and find an answer to that crucial question. This book has been an attempt to lay out the elements of that way, through the understanding of the molecular basis of life, its ability, from single cells to humans, to survive and even thrive in a planetary system like our Solar System, and finally what means we need to harness our human impulse for exploration to the pursuit of the search for life on other planets.

Perhaps the discovery of life on other worlds will encourage us, through humility, to see our own world in a more concerned way and impress on us the ethical and moral imperative to manage our planet in a better way than we have up to now.

Appendix 1
Bibliography

PART I

de Duve, C. (1995). *Vital Dust*. New York: Basic Books.

CHAPTER I

Balsiger, H., Beckers, J., Carusi, A., Geiss, J., Grieger, G., Horneck, G., Hyong, P., Jacob, M., Langevin, Y., Lemaire, J., Lequeux, J., Pinet, P., Spohn, T. and Wänke, H. (1992). *Mission to the Moon: Report of the Lunar Study Steering Group*. ESA SP-1150. Noordwijk: ESA-ESTEC.

Brack, A., Fitton, B. and Raulin, F. (1999). *Exobiology in the Solar System and the Search for Life on Mars*. ESA SP-1231. Noordwijk: ESA-ESTEC.

Darwin, C. (1859). *The Origin of Species*. London: John Murray.

DeBakey, M. E. (2000). Foreword. In *Challenges of Human Space Exploration*, ed. M. Freeman. Chichester: Springer and Praxis.

Horneck, G., Facius, R., Reichert, M., Rettberg, P., Seboldt, W., Manzey, D., Comet, B., Maillet, A., Preiss, H., Schauer, L., Dussap, C. G., Poughon, L., Belyavin, A., Reitz, G., Baumstark-Khan, C. and Gerzer, R. (2003). *HUMEX, a Study on the Survivability and Adaptation of Humans to Long-Duration Exploratory Missions*. ESA SP-1264. Noordwijk: ESA-ESTEC.

Innes, H. (1978). *The Last Voyage: Captain Cook's Lost Diary*. London: Collins.

Johnston, R. S., Dietlein, L. F. and Berry, C. A. (1975). *Biomedical Results of Apollo*. NASA SP 368. Washington, DC: NASA

McKay, C. P., Toon, O. B. and Kasting, J. F. (1991). Making Mars habitable. *Nature*, **352**, 489–96.

Moorehead, A. (1969). *Darwin and the Beagle*. Harmondsworth: Penguin Books.

Reichert, M., Seboldt, W., Leipold, M., Klimke, M., Fribourg, C. and Novara, M. (1999). *Promising Concepts for a First Manned Mars Mission*. ESA Workshop on 'Space Exploration and Resources Exploitation' (ExploSpace), Cagliari, Sardinia, Italy, October 1998. ESA Report WPP-151. Noordwijk: ESA-ESTEC.

Verne, J. (1993). *From the Earth to the Moon*. New York: Bantam.

Westall, F., Brack, A., Hofmann, B., Horneck, G., Kurat, G., Maxwell, J., Ori, G. G., Pillinger, C., Raulin, F., Thomas, N., Fitton, B., Clancy, P., Prieur, D. and

Details of the Amundsen, Scott and Shackleton expeditions may be found at www. South-pole.com.

Vassaux, D. (2000). An ESA study for the search for life on Mars. *Planet. Space Sci.*, **48**, 181–202.

PART II

Kauffmann, S. (1995). *At Home in the Universe.* New York: Oxford University Press, p. 31.

CHAPTER 2

Bailey, J. (2001). Astronomical sources of circularly polarized light and the origin of homochirality. *Origins Life Evol. Biosphere*, **31**, 167–83.

Bailey, J., Chrysostomou, A., Hough, J. H., Gledhill, T. M., McCall, A., Clark, S., Ménard, F. and Tamura, M. (1998). Circular polarization in star formation regions: implications for biomolecular homochirality. *Science*, **281**, 672–4.

Ban, N., Nissen, P., Hansen, J., Moore, P. B. and Steitz, T. A. (2000). The complete atomic structure of the large ribosomal subunit at 2.4 resolution. *Science*, **289**, 905–20.

Barbier, B. and Brack, A. (1992). Conformation-controlled hydrolysis of polyribonucleotides by sequential basic polypeptides. *J. Am. Chem. Soc.*, **114**, 3511–15.

Barbier, B., Henin, O., Boillot, F., Chabin, A., Chaput, D. and Brack, A. (2002). Exposure of amino acids and derivatives in the Earth orbit. *Planet. Space Sci.*, **50**, 353–9.

Bernal, J. D. (1949). The physical basis of life. *Proc. Roy. Soc. London*, **357A**, 537–58.

Bertrand, M., Bure, C., Fleury, F. and Brack, A. (2001). Prebiotic polymerisation of amino acid thioesters on mineral surfaces. In *Geochemistry and the Origin of Life*, ed. S. Nakashima, S. Maruyama, A. Brack and B. F. Windley Tokyo: Universal Academy Press, pp. 51–60.

Boillot, F., Chabin, A., Buré, C., Venet, M., Belsky, A., Bertrand-Urbaniak, M., Delmas, A., Brack, A. and Barbier, B. (2002). The Perseus exobiology mission on MIR: behaviour of amino acids and peptides in Earth orbit. *Origins Life Evol. Biosphere*, **32**, 359–85.

Brack, A. (1990). Extraterrestrial organic molecules and the emergence of life on Earth. In *Fourth European Symposium on Life Sciences in Space*, ESA SP-307, pp. 565–9.

Brack, A. (1993). From amino acids to prebiotic active peptides: a chemical restitution. *Pure & Appl. Chem.*, **65**(6), 1143–51.

Brack, A. and Spach, G. (1979). β-structures of polypeptides with L- and D-residues. Part I: Synthesis and conformational studies. *J. Molec. Evolution*, **13**, 35–46.

Burmeister, J. (1998). Self-replication and autocatalysis. In *The Molecular Origins of Life: Assembling Pieces of the Puzzle*, ed. A. Brack. Cambridge: Cambridge University Press, pp. 295–312.

Cooper, G., Kimmich, N., Belisle, W., Sarinana, J., Brabham, K. and Garrel, L. (2001). Carbonaceous meteorites as a source of sugar-related organic compounds for the early Earth. *Nature*, **414**, 879–83.

Cronin, J. R., Pizarello, S. and Cruikshank, D. P. (1988). Organic matter in carbonaceous chondrites, planetary satellites, asteroids, and comets. In *Meteorites and the Early Solar System*, ed. J. F. Kerridge and M. S. Matthews. Tucson, AZ: University of Arizona Press, pp. 819–57.

de Duve, C. (1998). Possible starts for primitive life. Clues from present-day biology: the thioester world. In *The Molecular Origins of Life: Assembling Pieces of the Puzzle*, ed. A. Brack. Cambridge: Cambridge University Press, pp. 219–36.

Deamer, D. W. (1985). Boundary structures are formed by organic components of the Murchison carbonaceous chondrite. *Nature*, **317**, 792–4.

Deamer, D. W. (1998). Membrane compartments in prebiotic evolution. In *The Molecular Origins of Life: Assembling Pieces of the Puzzle*, ed. A. Brack. Cambridge: Cambridge University Press, pp. 189–205.

Delsemme, A. (1998). Cosmic origin of the biosphere. In *The Molecular Origins of Life: Assembling Pieces of the Puzzle*, ed. A. Brack. Cambridge: Cambridge University Press, pp. 100–18.

Ehrenfreund, P. and Charnley, S. B. (2000). Organic molecules in the interstellar medium, comets and meteorites. *Ann. Review Astron. Astrophys.*, **38**, 427–83.

Ehrenfreund, P. *et al.* (2002). Astrophysical and astrochemical insights into the origin of life. *Rep. Prog. Phys.*, **65**, 1427–87.

Eschenmoser, A. (1999). Chemical etiology of nucleic acid structure. *Science*, **284**, 2118–24.

Ferris, J. P., Hill, Jr., A. R., Liu, R. and Orgel, L. E. (1996). Synthesis of long prebiotic oligomers on mineral surfaces. *Nature*, **381**, 59–61.

Fox, S. W. and Dose, Keds. (1977). *Molecular Evolution and the Origin of Life* (revised edition). New York: Marcel Dekker.

Haldane, J. B. S. (1929). The origin of life. *Rationalist Annual*, **148**, 3–10.

Holm, N. G. and Andersson, E. M. (1995). Abiotic synthesis of organic compounds under the conditions of submarine hydrothermal systems: a perspective. *Planet. Space Sci.*, **43**, 153–9.

Holm, N. G. and Charlou, J.-L. (2001). Initial indications of abiotic formation of hydrocarbons in the Rainbow ultramafic hydrothermal system, Mid-Atlantic Ridge. *Earth Planet. Sci. Lett.*, **191**, 1–8.

Huber, C. and Wächtershäuser, G. (1997). Activated acetic acid by carbon fixation on (Fe, Ni)S under primordial conditions. *Science*, **276**, 245–7.

Huber, C. and Wächtershäuser, G. (1998). Peptides by activation of amino acids with CO on (Ni, Fe) surfaces: implications for the origin of life. *Science*, **281**, 670–2.

Inoue, T. and Orgel, L. E. (1982). Oligomerization of guanosine 5′-phosphor 2-methylimidazolide on poly(C). An RNA polymerase model. *J. Mol. Biol.*, **162**, 204–17.

James, K. D. and Ellington, A. D. (1995). The search for missing links between self-replicating nucleic acids and the RNA world. *Origins Life Evol. Biosphere*, **25**, 515–30.

James, K. D. and Ellington, A. D. (1998). Catalysis in the RNA world. In *The Molecular Origins of Life: Assembling Pieces of the Puzzle*, ed. A. Brack. Cambridge: Cambridge University Press, pp. 269–94.

Joyce, G. F., Schwartz, A. W., Miller, S. L. and Orgel, L. E. (1987). The case for an ancestral genetic system involving simple analogs of the nucleotides. *Proc. Nat. Acad. Sci. USA*, **84**, 4398–402.

Kauffmann, S. (2002). In *The Next Fifty Years*, ed. by J. Brockman. New York: Vintage Books, p. 126.

Luisi, P. L. (1998) About various definitions of life. *Origins Life Evol. Biosphere*, **28**, 613–22.

Maurette, M. (1998). Carbonaceous micrometeorites and the origin of life. *Origins Life Evol. Biosphere*, **28**, 385–412.

Maurette, M., Duprat, J., Engrand, C., Gounelle, M., Kurat, G., Matrajt, G. and Toppani, A. (2000). Accretion of neon, organics, CO_2, nitrogen and water from large interplanetary dust particles on the early Earth. *Planet. Space Sci.*, **48**, 1117–37.

Miller, S. L. (1953). The production of amino acids under possible primitive Earth conditions. *Science*, **117**, 528–9.

Muñoz Caro, G. M., Meierhenrich, U. J., Schutte, W. A., Barbier, B., Arcones Segovia, A., Rosenbauer, H., Thiemann, W. H.-P., Brack, A. and Greenberg, J. M. (2002). Amino acids from ultraviolet irradiation of interstellar ice analogues. *Nature*, **416**, 403–6.

Nielsen, P. E., Egholm, M., Berg, R. H. and Buchardt, O. (1991). Sequence-selective recognition of DNA by strand displacement with a thymine-substituted polyamide. *Science*, **254**, 1497–500.

Oparin, A. I. (1924). *Proikhozndenie Zhizni*. Moscow: Moskowski Rabochi. See *The Origin of Life*. New York: Dover (1952).

Ourisson, G. and Nakatani, Y. (1994). The terpenoid theory of the origin of cellular life: the evolution of terpenoids to cholesterol. *Chemistry and Biology*, **1**, 11–23.

Paecht-Horowitz, M., Berger, J. and Katchalsky, A. (1970). Prebiotic synthesis of polypeptides by heterogeneous polycondensation of amino-acid adenylates. *Nature*, **228**, 636–9.

Palyi, G., Zucchi, C. and Caglioti L. eds. (2002). *Fundamentals of Life*. Paris: Editions Scientifiques et Médicales Elsevier.

Pizzarello, S. and Cronin, J. R. (2000). Non-racemic amino-acids in the Murray and Murchison meteorites. *Geochim. Cosmochim. Acta*, **64**(2), 329–38.

Schmidt, J. G., Nielsen, P. E. and Orgel, L. E. (1997). Information transfer from peptide nucleic acids to RNA by template-directed syntheses. *Nucleic Acids Res.*, **25**, 4797–802.

Schöning, K.-U., Scholz, P., Guntha, S., Wu, X., Krishnamurthy, R. and Eschenmoser, A. (2000). Chemical etiology of nucleic acid structure: the α-threofuranosyl-(3′→2′) oligonucleotide system. *Science*, **290**, 1347–51.

Schopf, J. W. (1993). Microfossils of the Early Archean Apex Chert: new evidence of the antiquity of life. *Science*, **260**, 640–6.

Schwartz, A. W. (1998). Origins of the RNA world. In *The Molecular Origins of Life: Assembling Pieces of the Puzzle*, ed. A. Brack. Cambridge: Cambridge University Press, pp. 237–54.

Schwartz, A. W. and Orgel, L. E. (1985). Template-directed synthesis of novel, nucleic acid-like structures. *Science*, **228**, 585–7.

Severin, K. S., Lee, D. H., Martinez, J. A. and Ghadiri, M. R. (1997). Peptide self-replication via template-directed ligation. *Chem. Eur. J.*, **3**, 1017–24.

Shibata, T., Yamamoto, J., Matsumoto, N., Yonekubo, S., Osanai, S. and Soai, K. (1998). Amplification of a slight enantiomeric imbalance in molecules based on asymmetry autocatalysis. *J. Amer. Chem. Soc.*, **120**, 12157–8.

Spach, G. (1984). Chiral versus chemical evolutions and the appearance of life. *Origins Life Evol. Biosphere*, **14**, 433–7.

Stoks, P. G. and Schwartz, A. W. (1982). Basic nitrogen-heterocyclic compounds in the Murchison meteorite. *Geochim. Cosmochim. Acta*, **46**, 309–15.

Terfort, A. and von Kiedrowski, G. (1992). Self-replication by condensation of 3-aminobenzamidines and 2-formyl-phenoxyacetic acids. *Angew. Chem. Int. Ed. Engl.*, **31**, 654–6.

Visscher, J. and Schwartz, A. W. (1989). Manganese-catalyzed oligomerizations of nucleotide analogs. *J. Mol. Evol.*, **29**, 284–7.

Visscher, J. and Schwartz, A. W. (1990). Oligomerization of cytosine-containing nucleotide analogs in aqueous solution. *J. Mol. Evol.*, **30**, 3–6.

Wächtershäuser, G. (1994). Life in a ligand sphere. *Proc. Natl. Acad. Sci. USA*, **91**, 4283–7.

Wächtershäuser, G. (1998). Origin of life in an iron-sulfur world. In *The Molecular Origins of Life: Assembling Pieces of the Puzzle*, ed. A. Brack. Cambridge: Cambridge University Press, pp. 206–18.

Wilde, S. A., Valley, J. W., Peck, W. H. and Graham, C. M. (2001). Evidence from detrital zircons for the existence of continental crust and oceans on the Earth 4.4 Gyr ago. *Nature*, **409**, 175–8.

Wintner, E. A., Conn, M. M. and Rebek, J. (1994). Studies in molecular replication. *Acc. Chem. Res.*, **27**, 198–203.

Yanagawa, H. and Kobayashi, K. (1992). An experimental approach to chemical evolution in submarine hydrothermal systems. *Origins. Life Evol. Biosphere*, **22**, 147–160.

Zaug, A. J. and Cech, T. R. (1986). The intervening sequence RNA of tetrahymena is an enzyme. *Science*, **231**, 470–5.

CHAPTER 3

Bernard, F. P., Connan, J. and Magot, M. (1992). Indigenous micro-organisms in connate water of many oil fields: a new tool in exploration and production techniques. *Soc. Petroleum Engineers*, **67**, 467–76.

Bernhardt, G., Luedemann, H. D., Jaenicke, R., Koenig, H. and Stetter, K. O. (1984). Biomolecules are unstable under black smoker conditions. *Naturwissenschaften*, **71**, 583–6.

Blochl, G., Rachel, R., Burgraff, S., Hafenbradl, D., Jannash, H. W. and Stetter, K. O. (1997). *Pyrolobus fumarii*, gen. and sp. nov., represents a novel group of archaea, extending the upper temperature limit for life to 113 °C. *Extremophiles*, **1**, 14–21.

Brack, A., Fitton, B. and Raulin, F. eds. (1999). *Exobiology in the Solar System and the Search for Life on Mars*. ESA SP-1231. Noordwijk: ESA-ESTEC.

Buecker, H. and Horneck, G. (1975). Studies on the effects of cosmic HZE-particles on different biological systems in the Biostack I and II flown on board of Apollo 16 and 17. In *Radiation Research*, ed. O. F. Nygaard, H. J. Adler and W. K. Sinclair. London: Academic Press, pp. 1138–51.

Cragg, B. A. and Parkes, R. J. (1994). Bacterial profiles in hydrothermally active deep sediment layers from Middle Valley (N. E. Pacific) Sites 857 and 858. *Proc. Ocean Drilling Prog. Sci. Results*, **139**, 509–16.

DeLong, E. F., Wu, K. Y., Prezelin, B. B. and Jovine R. V. M. (1994). High abundance of Archaea in Antarctic marine picoplankton. *Nature*, **371**, 695–7.

Driessen, A. J. M., Van de Vossenberg, J. L. C. M. and Konings, W. N. (1996). Membrane composition and ion-permeability in extremophiles. *FEMSMicrobiol. Rev.*, **18**, 139–48.

Forterre, P. (1995). Thermoreduction, a hypothesis for the origin of procaryotes. *Comptes Rendus de l'Acad. Sci. (Paris)*, **318**, 415–22.

Forterre, P. (1996). A hot topic: the origin of hyperthermophiles. *Cell*, **85**, 789–92.

Forterre, P. (1997). Protein versus rRNA: rooting the universal tree of life? *ASM News 63*, **2**, 89–95.

Furnes, H., Thorseth, I. H., Tumyr, O., Torsvick, T. and Fisk, M. R. (1996). Microbial activity in the alteration of glass from pillow lavas from hole 896A. *Proc. Ocean Drilling Prog. Sci. Results*, **148**, 191–206.

Galinski, E. A. and Tyndall, B. J. (1992). Biotechnological prospects for halophiles and halotolerant micro-organisms. In *Molecular Biology and Biotechnology of Extremophiles*, ed. R. A. Herbert and R. J. Sharp. London: Blackie, pp. 76–114.

Gold, T. (1992). The deep, hot biosphere. *Proc. Natl. Acad. Sci.*, **89**, 6045–9.

Grant, W. D. and Horikoshi, K. (1992). Alkaliphiles: ecology and biotechnological applications. In *Molecular Biology and Biotechnology of Extremophiles*, ed. R. A. Herbert and R. J. Sharp. London: Blackie, pp. 143–160.

Haridon, S. L., Reysenbach, A., Glenat, P., Prieur, D. and Jeanthon, C. (1995). Hot subterranean biosphere in continental oil reservoir. *Nature*, **377**, 223–4.

Horneck, G. (1992). Radiobiological experiments in space: a review. *Nucl. Tracks Radiat. Meas.*, **20**, 185–205.

Horneck, G. (1993). Responses of *Bacillus subtilis* spores to space environment: results from experiments in space. *Origins of Life*, **2**, 37–52.

Horneck, G., Buecker, H. and Reitz, G. (1994). Long-term survival of bacterial spores in space. *Adv. Space Res.*, **14**, (10)41–(10)45.

Horneck, G., Rettberg, P., Rabbow, E., Strauch, W., Seckmeyer, G., Facius, R., Reitz, G., Strauch, K. and Schott, J. U. (1996). Biological dosimetry of solar radiation for different simulated ozone column thicknesses. *J. Photochem. Photobiol. B: Biol.*, **55**, 389.

Horneck, G. and Baumstark-Khan, C. eds. (2000). *Astrobiology: the Quest for the Conditions of Life*. Berlin: Springer.

Keller, M., Braun, F.-J., Dirmeier, R., Hafenbradl, D., Burgrraf, S., Rachel, R. and Stetter, K. O. (1995). *Thermococcus alcaliphilus* sp. nov. a new hyperthermophilic archaeum growing on polysulfide at alkaline pH. *Arch. Microbiol.*, **164**, 390–5.

Kennedy, M. J., Reader, S. L. and Swierczynski, L. M. (1994). Preservation records of micro-organisms: evidence of the tenacity of life. *Microbiol.*, **140**, 2513–29.

Lindhal, T. (1993). Instability and decay of the primary structure of DNA. *Nature*, **362**, 709–15.

Marguet, E. and Forterre, P. (1994). DNA stability at temperatures typical for hyperthermophiles. *Nucl. Acid Res.*, **22**, 1681–6.

Nienow, J. A. and Friedman, E. I. (1993). Terrestrial lithophytic (rock) communities. In *Antarctica Microbiol.*, ed. E. Friedmann. New Jersey: Wiley, pp. 343–412.

Norris, P. R. and Ingledew, W. J. (1992). Acidophilic bacteria: adaptations and applications. In *Molecular Biology and Biotechnology of Extremophiles*, ed. R. A. Herbert and R. J. Sharp. London: Blackie, pp. 115–39.

Palmer, J. D. (1997). Organelle genomes: going, going, gone. *Science*, **275**, 790–1.

Parkes, R. J. and Maxwell, J. R. (1993). Some like it hot (and oily). *Nature*, **365**, 694–5.

Parkes, R. J., Cragg, B. A., Bale, S. K. *et al.* (1994). Deep bacterial biosphere in Pacific Ocean sediments. *Nature*, **371**, 410–3.

Prieur, D. (1992). Physiology and biotechnological potential of deep-sea bacteria. In *Molecular Biology and Biotechnology of Extremophiles*, ed. R. A. Herbert and R. J. Sharp. London: Blackie, pp. 163–97.

Reanney, D. C. (1974). On the origin of procaryotes. *J. Theor. Biol.*, **48**, 243–51.

Rueter, P., Rabus, R., Wilkes, H., Aeckersberg, F., Rainey, F. A., Jannasch, H. W. and Widdel, F. (1994). Anaerobic oxidation of hydrocarbons in crude oil by new types of sulphate-reducing bacteria. *Nature*, **372**, 455–8.

Russel, N. J. (1992). Physiology and molecular biology of psychrophilic micro-organisms. In *Molecular Biology and Biotechnology of Extremophiles*, ed. R. A. Herbert and R. J. Sharp. London: Blackie, pp. 203–21.

Schleper, C., Puehler, G., Holz, I., Gambacorta, A., Janekovic, D., Santarius, U., Klenk, H. P. and Zillig, W. (1995). *Picrophilus*: a novel aerobic, heterotrophic, thermoacidophilic genus and family comprising Archaea capable of growth around pH 0. *J. Bacteriol.*, **177**, 7050–9.

Smith, M. D., Masters, C. I. and Moseley, B. E. B (1992). Molecular biology of radiation resistant bacteria. In *Molecular Biology and Biotechnology of Extremophiles*, ed. R. A. Herbert and R. J. Sharp. London: Blackie, pp. 258–77.

Stetter, K. O. (1982). Ultra-thin mycelia-forming organisms from submarine volcanic areas having an optimum growth temperature of 105 °C. *Nature*, **300**, 258–60.

Stetter, K. O. (1996). Hyperthermophilic procaryotes. *FEMS Micibiol. Reviews*, **18**, 149–58.

Stetter, K. O., Huber, R., Blochl, E., Kurr, M., Eden, R. D., Fielder, M., Cash, H. and Vance, I. (1993). Hyperthermophilic archaea are thriving in deep North Sea and Alaskan oil reservoirs. *Nature*, **365**, 743–5.

Stevens, T. O. and McKinley, J. P. (1995). Lithoautotrophic microbial ecosystems in deep basalt aquifer. *Science*, **270**, 450–4.

Trent, J. D., Chastian, R. A. and Yayanos, A. A. (1984). Possible artefactual basis for apparent bacterial growth at 250 °C. *Nature*, **307**, 737–40.

Weber, P. and Greenberg, J. M. (1985). Can spores survive in interstellar space? *Nature*, **316**, 403–7.

CHAPTER 4

Arrhenius, S. (1903). Die Verbreitung des Lebens im Weltenraum. *Die Umschau*, **7**, 481–5.

Bischoff, A. and Stöffler, D. (1992). Shock metamorphism as a fundamental process in the evolution of planetary bodies: information from meteorites. *Europ. J. Mineral.*, **4**, 707–55.

Burchell, M. J., Mann, J., Bunch, A. W. and Brandao, P. (2001). Survivability of bacteria in hypervelocity impact. *Icarus*, **154**, 545–7.

Clark, B. C. (2001). Planetary interchange of bioactive material: probability factors and implications. *Origins Life Evol. Biosphere*, **31**, 185–97.

Clayton, R. N. and Mayeda, T. K. (1983). Oxygen isotopes in eucrites, shergotites, nakhlites and chassignites. *Earth and Planetary Science Letters*, **62**, 1–6.

Crick, F. H. C. (1981). *Life Itself, Its Origin and Nature.* New York: Simon and Schuster.

Darwin, C. (1871). Letter to Hocker, February 1871. In *Evolution on Planet Earth*, ed. L. J. Rothschild and A. M. Lister. London: Academic Press (2003), p. 110.

de Duve, C. (1995). *Vital Dust: Life as a Cosmic Imperative.* New York: Basic Books.

Dose, K., Bieger-Dose, A., Dillmann, R., Gill, M., Kerz, O., Klein, A., Meinert, H., Nawroth, T., Risi, S. and Stridde, C. (1995). ERA-experiment: 'space biochemistry'. *Adv. Space Res.*, **16**, (8)119–(8)129.

Friedmann, E. I. (1982). Endolithic micro-organism in the Antarctic cold desert. *Science*, **215**, 1045–53.

Haldane, J. B. S. (1928). The origin of life. *Rationalist Annual*, **148**, 3.

Horneck, G. (1992). Radiobiological experiments in space: a review. *Nucl. Tracks Radiat. Meas.*, **20**, 185–205.

Horneck, G. (1993). Responses of *Bacillus subtilis* spores to space environment: results from experiments in space. *Origins Life Evol. Biosphere*, **23**, 37–52.

Horneck, G. (2000). The microbial world and the case for Mars. *Planet. Space Sci.*, **48**, 1053–63.

Horneck, G. and Baumstark-Khan, C. eds. (2002). *Astrobiology, the Quest for the Conditions of Life.* Berlin: Springer.

Horneck, G. and Brack, A. (1992). Study of the origin, evolution and distribution of life with emphasis on exobiology experiments in Earth orbit. In S. L. Bonting, ed., *Advances in Space Biology and Medicine*, Vol. 2. Greenwich, CT: JAI Press, pp. 229–62.

Horneck G., Mileikowsky, C., Melosh, H. J., Wilson, J. W., Cucinotta, F. A. and Gladman, B. (2001). Viable transfer of micro-organism in the solar system and beyond. In G. Horneck and C. Baumstark-Khan eds. *Astrobiology, the Quest for the Conditions of Life.* Berlin: Springer, pp. 57–76.

Horneck, G., Rettberg, P., Reitz, G., Wehner, J., Eschweiler, U., Strauch, K., Panitz, C., Starke, V. and Baumstark-Khan, C. (2001). Protection of bacterial spores in space, a contribution to the discussion on Panspermia. *Origin Life Evol. Biosphere*, **31**, 527–47.

Horneck, G., Stöffler, D., Eschweiler, U. and Hornemann, U. (2001). Bacterial spores survive simulated meteorite impact. *Icarus*, **149**, 285–90.

Mancinelli, R. L., White, M. R. and Rothschild, L. J. (1998). Biopan Survival I: exposure of the osmophiles *Synechococcus* sp. (Nageli) and *Haloarcula* sp. to the space environment. *Adv. Space Res.*, **22**, (3)327–(3)334.

Mastrapa, R. M. F., Glanzberg, H., Head, J. N., Melosh, H. J. and Nicholson, W. L. (2000). Survival of *Bacillus subtilis* spores and *Deinococcus radiodurans* cells exposed to the extreme acceleration and shock predicted during planetary ejection. *Lunar Planet. Sci.*, **XXXI**. Abstract CD, 31st Lunar and Planetary Science Conference, 13–17 March 2000, Houston, TX.

Mattimore, V. and Battista, J. R. (1996). Radioresistance of *Deinococcus radiodurans*: functions necessary to survive ionizing radiation are also necessary to survive prolonged desiccation. *J. Bacteriol.*, **178**, 633–7.

McKay, C. P., Toon, O. B. and Kasting, J. F. (1991). Making Mars habitable. *Nature*, **352**, 489–96.

McKay, D. S., Gibson, E. K., Thomas-Kerpta, K. L., Vali, H., Romanek, C. S., Clemett, S. J., Chellier, X. D. F., Maechling, C. R. and Zare, R. N. (1996). Search for past life on Mars: possible relic biogenic activity in Martian meteorite ALH84001. *Science*, **273**, 924–30.

Melosh, H. J. (1985). Ejection of rock fragments from planetary bodies. *Geology*, **13**, 144–8.

Mileikowsky, C., Cucinotta, F. A., Wilson, J. W., Gladman, B., Horneck, G., Lindegren, L., Melosh. J., Rickman, H., Valtonen, M. and Zheng, J. Q. (2000). Natural transfer of viable microbes in space. Part 1: From Mars to Earth and Earth to Mars. *Icarus*, **145**, 391–427.

Miller., S. L. (1953). The production of amino acids under possible primitive Earth conditions. *Science*, **117**, 528.

NASA (2002). *A draft test protocol for detecting possible biohazards in Martian samples returned to Earth*. J. Rummel, M. S. Race, D. L. DeVincenzi, P. J. Schad, P. D. Stabekis, M. Viso and S. E. Acevedo, eds. NASA/CP-2002-211842. Moffett Field, CA: NASA ARC.

Nicholson, W. L., Munakata, N., Horneck, G., Melosh, H. J. and Setlow, P. (2000). Resistance of *Bacillus* endospores to extreme terrestrial and extraterrestrial environments. *Microb. Mol. Biol. Rev.*, **64**, 548–72.

Nicholson, W. L. (2003). Using thermal inactivation kinetics to calculate the probability of extreme spore longevity: implications for paleomicrobiology and lithopanspermia. *Origin Life Evol. Biosphere*, **33**, 621–31.

Nussinov, M. D. and Lysenko, S. V. (1983). Cosmic vacuum prevents radiopanspermia. *Origins of Life*, **13**, 153–64.

Oparin, A. I. (1924). *Proikhozndenie Zhizni*. Moscow: Moscowski Rabochi. See *The Origin of Life*. New York: Dover (1952).

Rothschild, L. J., Giver, L. J., White, M. R. and Mancinelli, R. L. (1994). Metabolic activity of micro-organisms in evaporites. *J. Phycol.*, **30**, 431–8.

Weiss, B. P., Kirschvink, J. L., Baudenbacher, F. J., Vali, H., Peters, N. T., MacDonald, F. A. and Wikswo, J. P. (2000). A low temperature transfer of ALH84001 from Mars to Earth, *Science*, **290**, 791–5.

CHAPTER 5

Abyzov, S. S. (1993). Micro-organisms in the Antarctic ice. In *Antarctic Microbiology*, ed. E. Friedmann, New York: Wiley & Sons, pp. 265–95.

Amann, R., Ludwig, W. and Schleifer, K.-H. (1992). Identification and *in situ* detection of individual bacterial cells. *FEMS Microbiol. Lett.*, **100**, 45–50.

Appel, P. W. U. and Moorbath, S. (1999). Exploring Earth's oldest geological record in Greenland. *Eos*, **80**, 257–64.

Barns, S. M., Fundyga, R. E., Jeffries, M. W. and Pace, N. R. (1994). Remarkable archaeal diversity detected in a Yellowstone National Park hot spring environment. *Proc. Natl. Acad. Sci. USA*, **91**, 1609–13.

Bartosch, S., Quader, H. and Bock. E. (1996). Confocal laser scanning microscopy: a new method for detecting micro-organisms in natural stone. *DECHEMA Monographs*, **133**, 37–43.

Biemann, K., Oro, J., Toulmin, P., Orgel, L. E., Nier, A. O., Anderson, D. M., Simmonds, P. G., Flory, D., Diaz, A. V., Rushneck, D. R., Biller, J. E. and Lafleur, A. L. (1977). Search for organic substances and inorganic volatile compounds in the surface of Mars. *J. Geophys. Res.*, **82**, 4641–58.

Burne, R. V. and Moore, L. S. (1987). Microbialites: organosedimentary deposits of benthic microbial communities. *Palaios*, **2**, 241–54.

Campbell, S. E. (1979). Soil stabilization by a procaryotic desert crust: implication for precambrian land biota. *Origins of Life*, **9**, 335–48.

Cano, R. J. and Borucki, M. K. (1995). Revival and identification of bacterial spores in 25- to 40-million-year-old Dominican amber. *Science*, **26**, 1060–4.

Carr, M. H. ed. (1996). *Water on Mars*. Oxford: Oxford University Press, p. 229.

Carr, M. H. and Wänke, H. (1992). Earth and Mars: water inventories as clues to accretional histories. *Icarus*, **98**, 61–71.

Chapelle, F. (1993). *Ground-Water Microbiology and Geochemistry*. New York: Wiley & Sons.

Deines, P. (1980). The isotopic composition of reduced organic carbon. In *Handbook of Environmental Isotope Geochemistry*, Vol.1, ed. P. Fritz and J. C. Fontes. Amsterdam: Elsevier, pp. 329–406.

Durand, B. ed. (1980). *Kerogen-Insoluble Organic Matter from Sedimentary Rocks*. Paris: Editions Techniq, p. 519.

Edwards, H. G. M., Russell, N. C., Weinstein, R. and Wynn-Williams, D. D. (1995). Fourier transform Raman spectroscopic study of fungi. *J. Raman Spectroscopy*, **26**, 911–16.

Eglinton, G. and Calvin, M. (1967). Chemical fossils. *Sci. Am.*, **216**, 32–43.

Friedmann, E. I. (1982). Endolithic micro-organisms in the Antarctic cold desert. *Science*, **215**, 1945–53.

Friedmann, E. I. ed. (1993). *Antarctic Microbiology*. New York: Wiley-Liss.

Gilichinsky, D. A., Soina, V. S. and Petrova, M. A. (1993). Cryoprotective properties of water in the Earth cryolithosphere and its role in exobiology. *Origins of Life*, **23**, 65–75.

Hayes, J. M., Kalpan, I. R. and Wedeking, K. W. (1983). Precambrian organic geochemistry: preservation of the record. In *Earth Earliest Biosphere: Its Origin and Evolution*, ed. J. W. Schopf. Princeton, NJ: Princeton University Press, pp. 93–134.

Hitchcock, D. R. and Lovelock, J. E. (1967). Life detection by atmospheric analysis. *Icarus*, 7, 149–59.

Horneck, G. (1995). Exobiology, the study of the origin, evolution and distribution of life within the context of cosmic evlolution: a review. *Planet. Space Sci.*, **43**, 189–217.

Horowitz, N. H., Hobby, G. L. and Hubbard, J. S. (1977). Viking on Mars: the carbon assimilation experiments. *J. Geophys. Res.*, **82**, 4559–662.

Imshenetsky, A. A., Lysenko, D. V. and Kazakov, G. A. (1978). Upper boundary of the biosphere. *Appl. Environm. Microbiol.*, **35**, 1–5.

Jannasch, H. W. and Mottl, M. J. (1985). Geomicrobiology and deep-sea hydrothermal vents. *Science*, **229**, 717–25.

Karl, D. M., Wirsen, C. O. and Jannasch, H. W. (1980). Deep-sea primary production at the Galapagos hydrothermal vents. *Science*, **20**, 1345–7.

Klein, H. P. (1979). The Viking mission and the search for life on Mars. *Rev. Geophys. Space Phys.*, **17**, 1655–62.

Kolbel-Boekel, J., Anders, E. and Nehrkorn, A. (1988). Microbial communities in the saturated groundwater environment. II. Diversity of bacterial communities in a Pleistocene sand aquifer and their in vitro activities. *Microbial. Ecology*, **16**, 31–48.

Levin, G. V. & Straat, P. A. (1977). Recent results from the Viking labeled release experiment on Mars. *J. Geophys. Res.*, **82**, 4663–7.

Lovelock, J. E. (1979). *Gaia*. Oxford: Oxford University Press.

MacDermott, A. J., Barron, L. D., Brack, A., Buhse, T., Drake, F., Emery, R., Gottarelli, G., Greenberg, J. M., Haberle, R., Hegstrom, R. A., Hobbs, K.,

Kondeputi, D.K., McKay, C., Moorbath, S., Raulin, F., Sandford, M., Schwartzman, D.W., Thiemann, W.H.-P., Tranter, G.E. and Zarnecki J.C. (1996). Homochirality as the signature of life: the SETH Cigar. *Planet. Space Sci.*, **44**, 1441–6.

McKay, C.P. (1986). Exobiology and future Mars missions: the search for Mars' earliest biosphere. *Adv. Space Res.*, **6**(12), 269–85.

McKay, C.P. (1991). Planetary evolution and the origin of life. *Icarus*, **91**, 93–100.

McKay, C.P. and Nedell, S.S. (1988). Are there carbonate deposits in the Valles Marineris, Mars? *Icarus*, **73**, 142–8.

McKay, C.P. and Stoker, C.R. (1989). The early environment and its evolution on Mars: implications for life. *Rev. Geophys.*, **27**, 189–214.

Mojzsis, S.J., Arrhenius, G., McKeegan, K.D., Harrison, T.M., Nutman, A.P. and Friend, R.L. (1996). Evidence for life on Earth before 3,800 million years ago. *Nature*, **384**, 55–9.

Moroz, V.I. and Mukhin, L.M. ed. (1978). *About the Initial Evolution of Atmosphere and Climate of the Earth-Type Planets*. Publication D-255. Moscow: Institute of Space Research, USSR Academy of Science, p. 44.

Nienow, J.A. and Friedmann, E.I. (1993). Terrestrial lithophytic (rock) communities. In *Antarctic Microbiology*, ed. E. Friedmann. New York: Wiley & Sons, pp. 343–412.

Owen, T., Biemann, K., Rushneck, D.R., Biller, J.E., Howarth, D.W. and Lafleur, A.L. (1977). The composition of the atmosphere at the surface of Mars. *J. Geophys Res.*, **82**, 4635–9.

Oyama, V.I., Berdahl, B.J. and Carle, G.C. (1977). Preliminary findings of the Viking gas exchange experiment and a model for Martian surface chemistry. *Nature*, **265**, 110–4.

Peters, K.E. and Moldowan, J.M. (1993). *The Biomarker Guide*. New York: Prentice Hall.

Pflug, H.D. (1978). Yeast-like microfossils detected in the oldest sediments of the Earth. *Naturwissenschaften*, **65**, 611–15.

Pflug, H.D. (1987). Chemical fossils in early minerals. *Topics in Current Chemistry*, **139**, 1–55.

Pollock, G.E., Miyamoto, A.K. and Oyama, V.I. (1970). The detection of optical asymmetry in biogenic molecules by gas chromatography for extraterrestrial space exploration: sample processing studies. *Life Sci. & Sp. Res.*, **8**, 9–107.

Rosing, M.T. (1999). ^{13}C depleted carbon microparticles in >3700 Ma seafloor sedimentary rocks from West Greenland. *Science*, **283**, 674–6.

Rothschild, L.J., Giver, L.J., White, M.R. and Mancinelli, R.L. (1994). Metabolic activity of micro-organisms in evaporites. *J. Physiol.*, **30**, 431–8.

Quader, H. and Bock, E. (1995). Konfokale-Laser-Raster-Mikroskopie: Eine neue Möglichkeit gesteinsbewohnende Mikroorganismen zu untersuchen. *Int. J. Restoration of Buildings*, **1**, 295–304.

Schidlowski, M. (1982). Content and isotopic composition of reduced carbon in sediments. In *Mineral Deposit and the Evolution of the Biosphere*, ed. H. D. Holland and M. Schidlowski. Berlin: Springer, pp. 103–22.

Schidlowski, M. (1988). A 3,800-million-year isotopic record of life from carbon in sedimentary rocks. *Nature*, **333**, 313–18.

Schidlowski, M. (1992). Stable carbon isotopes: possible clues to early life on Mars. *Adv. Space Res.*, **12**(4), 101–10.

Schidlowski, M. (1993a). The initiation of biological processes on Earth: summary of empirical evidence. In *Organic Geochemistry*, ed. M. H. Engel and S. A. Macko. New York: Plenum Press, pp. 639–55.

Schidlowski, M. (1993b). The beginnings of life on Earth: evidence from geological record. In *The Chemistry of Life's Origin*, ed. J. M. Greenberg *et al.* Deventer, The Netherlands: Kluwer Academic, pp. 389–414.

Schidlowski, M., Appel, P. W. V., Eichmann, R. and Junge, C. E. (1979). Carbon isotope geochemistry of the 3.7×10^9 yr old Ishua sediments, West Greenland: implications for the Archaean carbon and oxygen cycles. *Geochim. Cosmochim. Acta*, **43**, 189–99.

Schidlowski, M., Hayes, J. M. and Kaplan, I. R. (1983). Isotopic inferences of ancient biochemistries: carbon, sulfur, hydrogen and nitrogen. In *Earth's Earliest Biosphere: Its Origin and Evolution*, ed. J. W. Schopf. Princeton, NJ: Princeton University Press, pp. 149–86.

Schopf, J. W. ed. (1983). *Earth's Earliest Biosphere: Its Origin and Evolution.* Princeton, NJ: Princeton University Press, pp. XXV, 543.

Schopf, J. W. (1993). Microfossils of the Early Archaean Apex Chert: new evidence of the antiquity of life. *Science*, **260**, 640–6.

Schwartz, D. E., Mancinelli, R. L. and Kaneshiro, E. S. (1992). The use of mineral crystals as biomarkers in the search for life on Mars. *Adv. Space Res.*, **12**, (4)117–(4)119.

Siebert, J. and Hirsch, P. (1988). Characterization of 15 selected coccal bacteria isolated from Antarctic rock and soil samples from the McMurdo-Dry Valleys (South Victoria land). *Polar Biol.*, 8, 37–44.

Squyres, S. W. *et al* (2004). The opportunity rover's Athena science investigation at the Meridiana planum, Mars. *Science*, **306**, 1709–14.

Stetter, K. O. (1996). Hyperthermophilic procaryotes. *FEMS Microbiol. Rev.*, **18**, 149–58.

Stevens, T. O. and McKinley, J. P. (1995). Lithoautotrophic microbial ecosystems in deep basalt aquifers. *Science*, **270**, 450–4.

Stolz, J. F. (1985). The microbial community at Laguna Figueroa, Baja California, Mexico: from miles to microns. *Origins of Life*, **15**, 347–52.

Stolz, J. F. and Margulis, L. (1984). The stratified microbial community at Laguna Figueroa, Baja California, Mexico: a possible model for Prephanaerozoic laminated microbial communities preserved in cherts. *Origins of Life*, **14**, 671–9.

Summons, R. E. and Powell, T. G. (1992). Hydrocarbon composition of the late proterozoic oils of the Siberian Platform: implications for the depositional environment of source rocks. In *Early Organic Evolution: Implications for Mineral and Energy Resources*, ed. M. Schidlowski, S. Golubic, M. M. Kimberley, D. M. McKiroy and P. A. Trudinger. Berlin: Springer, pp. 296–307.

Summons, R. E., Jahnke, L. L. and Simoneit, B. R. T. (1996). Lipid biomarkers for bacterial ecosystems: studies of cultured organisms, hydrothermal environments and ancient sediments. In *Evolution of Hydrothermal Ecosystems on Earth. CIBA Foundation Symposium* 202, ed. G. R. Bock and J. A. Goode. Chichester: Wiley & Sons, pp. 174–93.

Vishniac, H. S. (1993). The microbiology of Antarctic soils. In *Antarctic Microbiology*, ed. E. Friedmann. New York: Wiley & Sons, pp. 297–341.

Walter, M. R. ed. (1976). *Stromatolites*. Amsterdam: Elsevier.

Walter, M. R. (1983). Archaean stromatolites: evidence of the Earth's earliest benthos. In *Earth's Earliest Biosphere: Its Origin and Evolution*, ed. J. W. Schopf. Princeton, NJ: Princeton University Press, pp. 187–213.

Weckwerth, G. and Schidlowski, M. (1995). Phosphorus as a potential guide in the search for extinct life on Mars. *Adv. Space Res.*, **15**(3), 185–91.

Westall, F., De Wit, M. J., Dann, J., Van Der Gaast, S., De Ronde, C. and Gerneke, D. (2001). Early Archaean fossil bacteria and biofilms in hydrothermally-influenced, shallow water sediments, Barberton greenstone belt, South Africa. *Precambrian Res.*, **106**, 93–116.

Woese, C. R. (1987). Bacterial evolution. *Microbiol. Rev.*, **51**, 221–71.

Wynn-Williams, D. D. (1988). Television image analysis of microbial communities in Antarctic fellfields. *Polarforschung*, **58**, 239–49.

Wynn-Williams, D. D. (1996). Response of pioneer soil microalgal colonists to environmental change in Antarctica. *Microb. Ecol.*, **31**, 177–88.

CHAPTER 6

Davies, P. (2002). In *The Next Fifty Years*, ed. J. Brockman. New York: Vintage Books, p. 165.

PART III

Mullane, R. Mike (1997). *Do Your Ears Pop in Space?* New York: Wiley. Nielsen, Dr. Jerrie (2001). *Ice Bound.* London: Ebury Press.

Lindsey, Clark S. (1999) to be found at www.hobbyspace.com/Active/controversy.

CHAPTER 7

Balsiger, H., Beckers, J., Carusi, A., Geiss, J., Grieger, G., Horneck, G., Hyong, P., Jacob, M., Langevin, Y., Lemaire, J., Lequeux, J., Pinet, P., Spohn, T. and Wänke, H. (1992). *Mission to the Moon: Report of the Lunar Study Steering Group.* ESA SP-1150. Noordwijk: ESA-ESTEC.

Brack, A., Fitton, B. and Raulin, F. (1999). *Exobiology in the Solar System and the Search for Life on Mars.* ESA SP-1231. Noordwijk: ESA-ESTEC.

Darwin, C. (1859). *The Origin of Species.* London: John Murray.

DeBakey, M. E. (2000). Foreword. In *Challenges of Human Space Exploration*, ed. M. Freeman. Chichester: Springer and Praxis.

DeVincenzi, D. L. (1992). Planetary protection issues and the future of exploration of Mars. *Adv. Space Res.*, **12**, (4)121–(4)128.

Gitelson J. I. and MacElroy, R. (1999). *Man-Made Closed Ecological Systems.* New York: Gordon & Breech.

Haynes R. H. and McKay C. P. (1992). The implantation of life on Mars: feasibility and motivation. *Adv. Space Res.*, **12**, (4)133–(4)140.

Horneck, G., Wynn-Williams, D. D., Mancinelli, R., Cadet, J., Munakata, N., Rontó, G., Edwards, H. G. M., Hock, B., Wänke, H., Reitz, G., Dachev, T., Häder D. P. and Brillouet, C. (1999). Biological experiments on the EXPOSE facility of the International Space Station. In *Proceedings of 2nd European Symposium on Utilisation of the International Space Station.* ESA SP-433, Noordwijk: ESA-ESTEC, pp. 459–68.

Horneck, G., Facius, R., Reitz, C., Rettberg, P., Baumstark-Khan, C. and Gerzer, R. (2001). Critical issues in connection with human planetary missions: protection of and from the environment. *Acta Astronautica*, **49**, 279–88.

Horneck, G., Facius, R., Reichert, M., Rettberg, P., Seboldt, W., Manzey, D., Comet, B., Maillet, A., Preiss, H., Schauer, L., Dussap, C. G., Poughon, L., Belyavin, A., Reitz, G., Baumstark-Khan, C. and Gerzer, R. (2003). *HUMEX, a Study on the Survivability and Adaptation of Humans to Long-Duration Exploratory Missions.* ESA SP-1264, Noordwijk: ESA-ESTEC.

Johnston R. S., Dietlein, L. F. and Berry, C. A. (1975). *Biomedical Results of Apollo.* NASA SP 368. Washington, DC: NASA.

Kelley, K. W. (1988). *The Home Planet*. Boston, MA: Addison Wesley.

McKay, C. P., Toon, O. B. and Kasting, J. F. (1991). Making Mars habitable. *Nature*, **352**, 489–96.

Moorehead, A. (1969). *Darwin and the Beagle*. Harmondsworth: Penguin Books.

Reichert, M., Seboldt, W., Leipold, M., Klimke, M., Fribourg, C. and Novara, M. (1999). *Promising Concepts for a First Manned Mars Mission*. ESA Workshop on 'Space Exploration and Resources Exploitation' (ExploSpace), Cagliari, Sardinia, Italy, October 1998. ESA Report WPP-151.

von Braun, W. (1953). *The Mars Project*. © Board of Trustees of the University of Illinois. Illini Books edition, 1991.

Westall, F., Brack, A., Hofmann, B., Horneck, G., Kurat, G., Maxwell, J., Ori, G. G., Pillinger, C., Raulin, F., Thomas, N., Fitton, B., Clancy, P., Prieur, D. and Vassaux, D. (2000). An ESA study for the search for life on Mars. *Planet. Space Sci.*, **48**, 181–202.

CHAPTER 8

Ryder, G. (1996). Humans and robots in the geological exploration of planets. *Ad. Astra*. **8**, 6 (Nov/Dec 1996), 20–1.

The roles of humans vs. robots in space exploration is dealt with at: www.hobbyspace.com/Active/controversy.html Humans vs. Robots.

Details on past human spaceflight experience can be found at: www.jsc.nasa.gov/history/hrf_history.htm.

Details on Skylab, Salyat and MIR space stations can be found at: en.wikipedia.org/wiki/Space_station.

CHAPTER 9

Ethical aspects of space have been discussed in a Groupe de reflexion 'Espace, Ethique et Société' convened by CNES, which published a report entitled *L'Espace et l'Ethique* (23 October 2000). A more elaborated development can be found in *La seconde chance d'Icare – Pour une éthique de l'espace*, by Jacques Arnould, Les Editions du Cerf, Paris, 2001.

DeVincenzi, D. L. and Stabekis, P. D. (1984). Revised planetary protection policy for solar system exploration. *Adv. Space Res.*, **4**, (12)291–(12)295.

McKay, C. P., Toon, O. B. and Kasting, J. F. (1991). Making Mars habitable. *Nature*, **352**, 489–96.

Rummel, J. D. (1989). Planetary protection policy overview and application to future missions. *Adv. Space Res.*, **9**, (6)181–(6)184.

Rummel, J. D. *et al.* (2002). Report of the COSPAR/IAU Workshop on Planetary Protection. Paris: COSPAR.

United Nations Treaties: Article IX, UN Doc. A/RES/2222/(XXI) 25 January 1967; TIAS No. 6347, 1967. Agreement (UN Gen. Ass. Resol. A/34/68, 5 December 1979)

PART IV

Dick, S. J. (1998). *Life on Other Worlds.* Cambridge: Cambridge University Press.

Guthke, Karl S. (1993). *The Last Frontier. Imaginary Other Words, from the Copernican Revolution to Modern Science Fiction.* Ithaca, NY: Cornell University Press.

Von Braun, W. (1953). *The Mars Project.* © Board of Trustees of the University of Illinois. Illini Books edition, 1991.

Zubrin, R. (1997). *The Case for Mars.* New York: Free Press.

CHAPTER 10

Details on propulsion technology can be found at: en.wikipedia.org/wiki/Rocket_propulsion.

Life support and *in-sister* exploitation technologies are dealt with in the ESA SP-1264 HUMEX report (see bibliography for Chapter 1).

CHAPTER 11

US National Academy of Sciences (2003). Report on *Frontiers in the Solar System: An Integrated Exploration Strategy.* Washington, DC: National Academies Press.

The past, present and future mission plans of NASA and ESA can be found at their respective websites: www.nasa.gov and www.esa.int.

The Louis Frank small comets issue can be found at http://smallcomets.physics.uiowa.edu/.

CHAPTER 13

Holland, H. J. (2002). In *The Next Fifty Years,* ed. J. Brockman. New York: Vintage Books, p. 181.

Lanier, J. (2002). In *The Next Fifty Years,* ed. J. Brockman. New York: Vintage Books, p. 217.

PART V

Cousteau, J-M. (1999). Quoted in the *Los Angeles Times.*

Crawford, I. A. (1998). The scientific case for human spaceflight. *Astronomy and geophysics,* **39**, 6.14–6.17.

CHAPTER 14

Dean, Darrick A. (1998). The Benefits and Necessity of Manned Exploration of Frontiers as Compared to Unmanned Efforts. American Society of Mechanical Engineers can be seen at: www.geocities/CapeCanaveral/Hangar/4264/Asme.html.

Dobbs, Lou (2001). *Space, the Next Business Frontier.* New York: Simon and Schuster.

McKnight, J. C. (2002). The Spacefaring Web 2.2: Critical Response. www.hobbyspace.com/AAdmin/archive/SpacefaringWeb/John_Carter_McKnight_-_2-02.html.

Snow, C. P. (1956). The two cultures and the scientific revolution. In *The Two Cultures.* Cambridge: Cambridge University Press.

Space controversies are dealt with well at: http://www.hobbyspace.com/Active/controversy.

Peter A. Taylor is available at murmur@ghg.net.

Woolley, R. (1956). Quoted in the *Daily Telegraph*, London, 3 January 1956.

Woolley, R. (1969). Interview in the *Daily Express*, London, 20 July 1969.

Appendix 2
Properties of water conducive to biology

When organic compounds form long chains they are called polymers. When such polymers are implicated in the chemistry of life they are called biopolymers. The main biopolymers in a cell, i.e. nucleic acids, proteins and membranes, contain C_xH_yO,N,S-groups and C_xH_y-groups (hydrocarbon groups). C_xH_yO,N,S-groups, especially those bearing groups which can easily be ionised such as -COOH or $-NH_2$, form hydrogen bonds with water molecules, are soluble in and have an affinity for water, i.e. are hydrophilic or water attracting. The large dipole moment of water (1.85 debye) favours the dissociation of the groups that ionise, while the high dielectric constant ($\epsilon = 80$) prevents the ions from recombining because the attractive forces for ion recombination are proportional to $1/\epsilon$. This is also true for salt ions that are associated with the biopolymers. C_xH_y-groups cannot form hydrogen bonds with water molecules and tend to avoid water molecules as much as possible. They are insoluble in water and hydrophobic. These two groups coexist in biopolymers and this coexistence leads to the emergence of favoured geometries of the biopolymers in water, such as α-helices, β-sheets, micelles, vesicles and liposomes mentioned above.

Water stabilises the biopolymer geometry by hydrophobic clustering. When mini-proteins based on an alternation of hydrophobic and hydrophilic amino acids are dissolved in water in the presence of salt, β-sheet bilayers are formed by clustering of the hydrophobic side-chains. When the properties of water are modified by adding increasing amounts of alcohol, the β-sheet structure is destroyed as a result of the relaxation of the water constraint on the hydrophobic clusters.

Due to β-structure formation, strictly alternating hydrophobic–hydrophilic sequences are stable over a wide temperature range and also resistant to chemical degradation.

Water can operate as a chemical reactant in many biochemical reactions. The most powerful example is given by oxygenic photosynthesis, the biological process by which plants and cyanobacteria use the solar energy to transform carbon dioxide into biological molecules. This reaction requires water according to the reaction

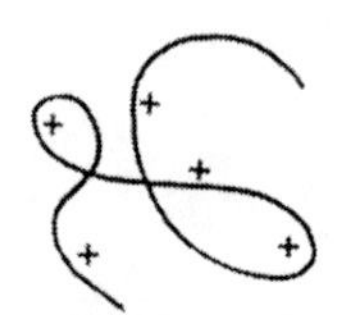

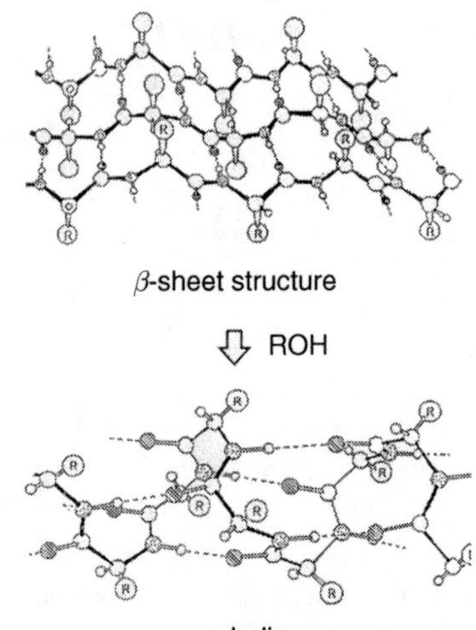

FIGURE A Geometries adopted by synthetic mini-proteins as a function of the environment. The mini-protein adopts a disordered geometry in pure water because of charge repulsions. When the charges are screened by salts, it undergoes a coil-to-sheet transition driven by the clustering of the hydrophobic groups. Addition of alcohol to the sheet releases the constraint of water and induces the formation of a helix.

$$CO_2 + 2\,H_2O + \text{light} \rightarrow [\text{CHOH}] + O_2 + H_2O$$

Water is also a powerful hydrolytic chemical reactant. Laboratory experiments have shown that water drives chemical reactions that would not occur in a non-aqueous context.

Water conducts heat easily and is therefore a good dissipater of heat. Hydrothermal systems that lie deep in the sea may be likely places for the synthesis of organic molecules that could be implicated in life. However, hydrothermal vents are often disregarded as efficient reactors for the synthesis of bio-organic molecules because of the high temperatures in them. However, the products that are synthesised in hot vents are rapidly quenched in the surrounding cold water thanks to the good heat conductivity of water. Short mini-proteins containing up to eight amino acids were obtained when a fluid containing amino acids was repeatedly circulated through the hot (225 °C) and cold (0 °C) regions in a laboratory reactor simulating a hydrothermal system; glycine peptides as complex as octaglycine were obtained.

Appendix 3
Why life may favour one-handedness

As we know, proteins are built up with 20 different amino acids and each amino acid, with the exception of glycine, exists in two mirror-image forms. Proteins in terrestrial life use only the L-form amino acids. It turns out that life that would simultaneously use both the right- and left-handed forms of the same biological molecules appears, in the first place, very unlikely for the following geometrical reasons.

Proteins adopt asymmetrical rigid geometries, right-handed α-helices and β-sheets, which play a key role in their catalytic activity. A. Brack and G. Spach in 1979 have shown that enzyme β-pleated sheets cannot form when both L- and D-amino acids are present in the same chain. Since the catalytic activity of an enzyme is intimately dependent upon the geometry of the chain, the absence of β-pleated sheets would impede, or at least considerably reduce, the potential range of activity of the enzymes. The use of one-handed biomonomers also reduces the complexity of the sequences. For a chain made of n units, the number of sequence combinations will be divided by $2^n/2$ when the system uses only one-handed (homochiral) monomers.

L-alanine-L-alanine-D-alanine
D-alanine-L-alanine-L-alanine
L-alanine-D-alanine-L-alanine
L-alanine-D-alanine-D-alanine
D-alanine-D-alanine-L-alanine
D-alanine-L-alanine-D-alanine
L-alanine-L-alanine-L-alanine
D-alanine-D-alanine-D-alanine

FIGURE B The number of sequences drops from 8, when both D and L enantiomers of amino acids (here alanine) are used, to 2 in the case of homochiral sequences.

Taking into account the fact that enzyme chains are generally made up of hundreds of monomers, and that nucleic acids contain several million nucleotides, the tremendous gain in simplicity offered by the use of monomers restricted to one-handedness is self evident. Thus, homochirality can be a crucial signature for life.

Note: The references for Appendix 3 may be found in the bibliography entry for Chapter 2 in Appendix 1.

Appendix 4
RNA analogues and surrogates in prebiotic chemistry

RNA analogues and surrogates have also been studied. The initial proposal by G. Spach in 1984 and G. F. Joyce and others in 1987 was that the first RNA was a flexible, non-handed derivative where ribose was replaced by glycerol, but these derivatives did not polymerise under prebiotic conditions. The ease of forming pyrophosphate (double phosphate) bonds prompted investigation of linking nucleotides by pyrophosphate groups. This proposal was tested using the reactions of the diphosphorimidazolides of deoxynucleotides by A. W. Schwartz and L. E. Orgel in 1985 and by J. Visscher and A. W. Schwartz in 1989 and 1990. The reaction in the presence of magnesium or manganese ion resulted in the formation of mini RNA analogues 10–20 nucleotides long.

Considering the facile formation of hexose-2,4,6-triphosphates from glycolaldehyde phosphate in a process analogous to the formose reaction, A. Eschenmoser and co-workers in 1999 chemically synthesised polynucleotides containing hexopyranose ribose (pyranosyl-RNA or p-RNA) in place of the usual 'natural' pentofuranose ribose found in RNA. p-RNAs form Watson-Crick-paired double helices that are more stable than RNA. Furthermore, the helices have only a weak twist, which should make it easier to separate strands during replication. Replication experiments have had marked success in terms of sequence copying but have failed to demonstrate turnover greater than one. p-RNA seems to be an excellent choice as a genetic system if it can be demonstrated that the prebiotic synthesis of pyranosyl nucleotides is much easier than synthesis of the standard isomers. The chemical synthesis of threo furanosyl nucleic acid, TNA, an RNA analogue built on the furanosyl form of the tetrose sugar threose, was also reported by K.-U. Schoning and others of the Eschenmoser group. TNA strands are much more stable than RNA to hydrolysis in aqueous solution. They form complementary duplexes between complementary strands but, of even greater potential importance, they form complementary strands with RNA. This raises the possibility that TNAs could have served as templates for the formation of complementary RNAs by template-directed synthesis. TNA is a more promising precursor to RNA than p-RNA because tetroses have the potential to be synthesised from glycolaldehyde phosphate and two other carbon precursors which may have been present in quantities greater than those of ribose on the primitive Earth.

Peptide nucleic acids (PNA) first synthesised by P. E. Nielsen and co-workers consist of a peptide-like backbone flanked by nucleic acid bases. PNAs form very stable double helical structures with complementary strands of PNA (PNA-PNA), with DNA (PNA-DNA) and with RNA (PNA-RNA) and even stable PNA_2-DNA triple helices. Information can be transferred from PNA to RNA, and vice versa, in template-directed reactions as described by J. G. Schmidt and others. Although PNA hydrolyses rather rapidly, thus restricting considerably the chances of PNA to have ever accumulated in the primitive terrestrial oceans, the PNA-PNA double helix illustrates that genetic information can be stored in a broad range of double helical structures.

Chemists are also tempted to think that primitive self-replicating systems must have used simpler informational molecules than biological nucleic acids or their analogues. Since self-replication is, by definition, autocatalysis, they are searching for simple autocatalytic molecules capable of selecting out favourable mutations. Different templates have been tested including mini-proteins by K. S. Severin and others, and other molecules by A. Terfort and G. von Kiedrowski in 1992 and E. A. Wintner and others in 1994. This has all been reviewed by J. Burmeister. In most cases, the rate of the autocatalytic growth did not vary in a linear sense. The initial rate of autocatalytic synthesis was found to be proportional to the square root of the template concentration, a limiting factor as compared to most autocatalytic reactions known so far. Autocatalytic reactions are particularly attractive since they might amplify small left/right or right/left excesses, even extraterrestrial, leading to full homochirality as suggested by T. Shibita and others.

Note: The references for Appendix 4 may be found in the bibliography entry for Chapter 2 in Appendix 1.

Appendix 5
Analysis techniques and technology for the detection of traces of extant or extinct life

EXTANT LIFE

Cellular versus non-biological structures

To differentiate between cellular structures and similar structures of non-biological origin, such as soil particles or aerosols, dyes have been used that specifically bind to subcellular components (e.g. acridine orange binds with cellular nucleic acids to form a fluorescent dye). When excited by UV light, the nucleic acid-acridine orange complex emits visible light which can be observed by fluorescence microscopy as described by F. Chapelle.

Recently, confocal laser scanning microscopy (CLSM) has been introduced to stone ecology by H. Quader, E. Bock and S. Bartosch to obtain a 3D visualisation of stone-inhabiting microbial communities. This CLSM method allows one to visualise micro-organisms in transparent stones such as quartzite over a depth of approximately 200 μm. CLSM gives a snapshot inside the stone matrix of the resident microbial community that consists of micro-organisms of different sizes and shapes occurring also in micro-aggregates or micro-colonies composed of from several up to hundreds of cells. To observe further minute details of the cells, which are not visible in light microscopy, such as cell walls, membranes and vacuoles, scanning electron microscopy (SEM) and transmission electron microscopy (TEM) have been applied.

These methods provide further details on the structure of subcellular components, e.g. Gram-positive or Gram-negative cell walls, and the occurrence of inclusions typical for a metabolic pathway, such as the energy-storing compound polybetahydroxybutyrate as described by F. Chapelle. Figure 5.1 shows a series of scanning electron micrographs of micro-organisms within ancient layers of the Antarctica ice sheet at depths of 1.6–2.4 m made by S. S. Abyzov.

Nucleic acid sequencing

Nucleic acid sequencing technology has provided a powerful tool for establishing the evolutionary relationships of organisms and for grouping micro-organisms according to their genotype, independent of any phenotypic observation. This is based on the fact that basic biochemical features, such as the genetic code, the set of amino acids in the proteins and the machinery of transcription of the genetic code and its translation into the proteins are common to all forms of life on Earth. During the evolution of life on Earth, at the level of the genotype, changes constantly occur randomly in time. C. R. Woese, by comparing the genome information stored in the nucleic acids and proteins, grouped all organisms into three domains of the phylogenetic tree: Bacteria, Archaea and Eucarya. As biological macromolecules, 16S rRNA, DNA-dependent polymerases and proton-pumping ATP-ases have been most commonly used. Different methods are applied to determine the percentage sequence similarity of the macromolecules of an unknown organism with that of a well-established representative of a domain.

However, even under terrestrial conditions, a large number of micro-organisms are not amenable to cultivation. Included are those from hot springs, where, however, the presence of a large number of hitherto unknown species has been identified by the DNA extracted from samples, which was then amplified and compared with known 16S rRNA sequences by S. M. Barns and others.

The viability of micro-organisms collected from extreme environments and their biomass can also be determined directly from biochemical analyses without cultivation of the cells. An example is the determination of the concentration of adenosine 5′-triphosphate (ATP) or of total adenylates in cells by D. M. Karl and others. A high concentration of ATP or total adenylates indicates a substantial population of viable micro-organisms.

As an alternative to the isolation of micro-organisms from sites under investigation, their *in-situ* examination in the undisturbed environment allows a determination of the extent and distribution of microbial colonisation. This has been achieved in the Antarctic fellfield soils by a combination of epifluorescence microscopy and television image analysis by the British Antarctic Survey's David Wynn-Williams. Using special staining techniques, the cells can be made visible relative to their opaque substrate. By using stains that selectively stain viable cells better than non-viable ones, the degree of viability of the population can be determined. Using this technique, the microalgal colonisation of the Antarctic soil surface was successfully recorded over a period of six years. R. Amann and others have used rRNA targeted nucleic acid probes to identify individual procaryotic cells in order to detect micro-organisms that are not amenable to cultivation. Such probes are either highly specific at the species/subspecies level or rather unspecific, reacting with all living organisms. They have been widely used to identify *in situ* the phylogeny of uncultured micro-organisms and to analyse complex microbial communities.

Optical handedness measurements

The optical handedness of monomers, e.g. amino acids in proteins and sugars in nucleic acids, is a conditio sine qua non for maintaining the secondary, tertiary and quaternary structure of the polymers and is therefore considered as a basic requirement for terrestrial life. Therefore, evidence for optical activity in a class of compounds that may make up polymers represents an interesting biomarker. G. E. Pollock and others have made measurements of optical activity using extracts by gas chromatography. Alexandra MacDermott and others have made similar measurements directly with the sample collected using polarised light and a light detector. A non-intrusive *in-situ* high-precision analysis of the distribution of organic and inorganic components of samples containing living micro-organisms is obtained by Fourier transform Raman spectroscopy. H. G. M. Edwards and others have described how, from the vibrational bands, the presence of organic compounds, such as chlorophyll and calcium oxalate di- and monohydrate, has been detected in different zones of endolithic communities.

EXTINCT LIFE

Biogeochemical evidence

Apart from microfossil relics, organisms also leave a chemical record of their former existence. While the bulk of the body material degrades after the death of the organism, with most of the carbon ending up in CO_2, a minuscule fraction of the organic substance (between 0.1 % and 0.01 %) usually escapes decomposition by burial in newly-formed sediments. Here, the primary biological polymers undergo a large-scale reconstitution with subsequent transformation into inorganic carbon polymers, resulting finally in the formation of kerogen, a chemically inert condensed aggregate of aliphatic and aromatic hydrocarbons that represents the end-product of the diagenetic alteration of biological matter in sediments as described by B. Durand. As the final residue of living substances, kerogenous materials and their graphitic derivatives constitute strong evidence of past biological activity.

A third category of biochemical evidence of ancient life is represented by quasi-pristine organic molecules (mostly pigments or single discrete hydrocarbon chains) that have preserved their identities during the evolution from dead organic matter to the polymerised aggregates of solid hydrocarbons that mark the end of that process. The record of such 'biomarker' molecules or 'chemofossils' has been described by G. Eglinton and M. Calvin and goes far back into Precambrian times. R. E. Summons and T. G. Powell have indicated the significance of this in connection with the geologically oldest (Proterozoic) petroleum occurrences.

Molecular biomarkers ('chemical fossils') in sediments

Biological markers, or 'biomarkers', are terms used for the wide variety of organic compounds that are derived from living organisms and can be found in sediments. Such compounds have also been termed 'chemical fossils' but the most commonly-used term is that of biomarker, as given by K. E. Peters and J. M. Moldowan. Biomarkers are extractable from sediments and fossils by organic solvents, the most studied being of the lipid type, although certain pigments, e.g. metal porphyrins, are also referred to as biomarkers. The key to the use of biomarkers is frequently their complexity of structure, since this represents a major inheritance of original biologically derived information. In addition, the molecular distributions in terms of their relative amounts can tell a lot, as the original biosynthesis imparts characteristic distributions of the different compounds. However, subsequent diagenetic alterations and mineralisation modify those patterns.

Molecular characteristics of the biosphere

What is the recognisable record of the past biosphere in the surface of the Earth? In general terms, the biosynthetic imprint of a living organism is seen in the carbon skeletons, the functionalities and the stereochemistries of the compounds it manufactures using the enzymatic, molecular machinery described in Chapter 2. The fundamental scheme of life's universal, but extremely precise, molecular machinery is simply:

DNA ====> RNA ====> Proteins ====> Metabolites

DNA and RNA consist of genetically determined sequences of five different nucleotide bases, with the DNA sequence determining the RNA sequence. It, in turn, controls the synthesis of proteins from some 20 homochiral alpha amino acids. The proteins consist of short (10–1000 amino acids) polypeptide chains and normally generate their own specific intramolecular crosslinks, thereby defining the precise shape and location of reactive functionalities required for enzymatic activity. Proteins in turn catalyse the synthesis of specific metabolites and homologous metabolic series. Enzymatic synthesis (metabolism) and breakdown (catabolism) of molecules provides for all the needs of the organisms. The metabolites comprise a large range of molecules that include, *inter alia*, the building blocks to synthesise a variety of cellular biochemicals, such as lipids, polysaccharides and co-enzymes.

Chiral centres such as those at a single tetrahedrally substituted carbon atom are almost always synthesised with a single chirality. Very unusually, an organism may make both, while, in other rare instances, one enantiometer may be made by one species and the other by a different species. Cholesterol, a sterol widely distributed in eucaryotes, is an example of a highly ordered carbon skeleton in which the overall shape is dependent upon the stereochemistry of the ring fusions and of the methyl and hydroxyl substituents of the rings. It has eight chiral centres. The structure is highly specific, both in the gross skeleton and in the stereochemistry, so that it makes an excellent biomarker, one which has been traced back hundreds of millions of years in immature sediments. It is also a carrier of abundant isotopic information, due to the biosynthesis pathway. Thus carbon isotope measurements can help to define the origin of the cholesterol found in such sediments, since its $\delta^{13}C$ value will reflect that of the carbon pool from which the individual C5 units were assembled.

Imprint of the biosphere in the geosphere

Studies of young immature sediments reveal evidence of decayed biological objects as microscopically recognisable organic debris, and molecular evidence as detectable concentrations of simple and complex biochemicals of both original and diagenetically modified structures. For example, amino acid analyses of teeth, bones and shells in young sediments display distributions characteristic of proteins which reveal increased racemisation with related temperature and time of burial, a measure that can even be used for dating purposes. Similar changes apply to other types of molecules, e.g. steroids, triterpenoids, alkaloids and the whole suite of secondary metabolites synthesised by the enzymes in living systems.

The initial record is very close to that of contemporary biochemistry, with small amounts of fairly well preserved DNA, protein and carbohydrates still detectable. Most of this initial record is rapidly removed within a few tens of years, principally by microbial action. However, removal and destruction is selective, patchy and incomplete. Thus, the end result is that relatively easily biodegraded compounds and biological polymers survive in small to trace amounts for thousands to millions of years in aquatic sediments that have not been deeply buried and subjected to geothermal heating. Some micro-environments, such as frozen sediments, anoxic sediments, organic resins, copal and amber, result in an exceptional preservation of biochemicals. Certainly, low temperature, the restriction of access afforded by the close packing of clay particles, absence of reactive species such as free radicals, lack of nutrients (e.g. electron donors such as nitrate, sulphate), and lack of liquid water all have a preservative role. At the molecular level, the fundamental factors facilitating preservation seem to be restrictions on the mobility of molecules and increased steric hindrance at reactive centres, both intra- and intermolecularly. All molecules are ultimately degradable, whether by micro-organisms, reagents and heat, operating over time, either singly or together. But differences in intramolecular bond strengths are fundamental controls on molecular persistence, especially in regard to non-biologically mediated reactions. Thus, novel aliphatic type and aromatic type biopolymers, whose skeletons are based essentially on carbon-carbon bonds, have been detected recently in several species of aquatic unicellular algae, and have been shown to persist in significant amounts in both recently deposited and ancient sediments. These biomaterials, or rather their alteration products, may make up a significant portion of sedimentary organic matter.

Bulk terrestrial organic matter

The bulk (about 98 %) of the organic matter held in the sedimentary rocks of the Earth's crust is the insoluble, amorphous, organic 'kerogen'. In young, immature sediments it is rich in hydrogen and contains other elements such as oxygen, sulphur and nitrogen in lesser amounts; while in old, mature sediments it is highly carbonaceous, even graphitic, and is very low in hydrogen and elements other than carbon. Whatever its state, kerogen is almost entirely derived ultimately from biological debris which, when immature, contains component pieces of biomacromolecules and smaller metabolites crosslinked into the kerogen structure by carbon-carbon, carbon-sulphur and other bonds. Some components may be trapped between the molecules, while others may have suffered additional modification through oxidation and other processes. Experiments specifically designed to release molecular moieties from this heteropolymetric material have shown that the kerogen matrix has protected some components from diagenetic and other changes experienced by free unbonded biomarkers in the same sediment. However, the original biosynthetic order incorporated as molecular components of the kerogen is gradually obscured and destroyed by the diagenetic processes in the sediment. Indeed, deep burial with accompanying rise in geothermal temperature, pressure and the passage of time eventually erases almost all of the original biosynthetic order as the carbonisation proceeds, e.g. the biosynthetic dominance of odd-over-even carbon numbers in the C_{30} region of plant wax alkanes is gradually replaced by a smooth distribution at lower carbon numbers, as the molecules are broken down through thermally-induced carbon-carbon bond breakage.

Evidence of Extant Life

1. **Structural Indications:**

   Microscopic Observation: of cells and subcellular structures. Size distributions, dividing cells and selective dyes to identify specific cellular components, for example, acridine orange can indicate nucleic acid.

Confocal Laser Scanning Microscopy: provides very high-resolution 3D imagery, including observation of communities inside transparent rocks.
Electron Microscope: to provide nm-resolution of cell walls, membranes, vacuoles, etc.
Macroscopic Observation: of large (10 mm) microbial communities in, for example, sandstone and evaporites. Desert crusts of cyanobacteria and mats on hypersaline lagoons may be cm thick.

2. **Culture Indicators:** isolation and successful culturing, with subsequent biochemical analysis, for example, of nucleic acid sequencing, protein, lipid and sugar content, provides the most unequivocal evidence of extant life.
3. **Metabolic Indicators:** the chemical products of metabolism, in particular the gaseous products, can be observed in culture or *in-situ* colonies, using isotopic tracers or direct analysis, e.g. gas chromatography. The Viking experiments were based on testing for carbon assimilation, catabolic activity and respiration.
4. **Isotopic Indicators:** the discrimination against ^{13}C relative to ^{12}C during enzymatic uptake of carbon in photosynthesis is a valuable biomarker, whose record extends back 3.5 Gyr. Combining compound-specific isotopic data with diagnostic structural information permits specific characterisation.
5. **Chirality Indicator:** homochirality is a characteristic of life; without it, polymerisation and template replication do not proceed effectively. Determination can be by observing optical activity.
6. **Spectral Observations:** vibrational spectroscopy, e.g. Raman spectroscopy, can detect a variety of organic compounds involved in living systems.

Evidence of Extinct Life

1. **Structural Indicators:**

   Microscopic Observations: of groups of possible microfossil structures in sedimentary deposits, including petrological analysis of the associated minerals and the study of any residual carbonaceous matter.
   Electron Microscopy: to provide nm-resolution study of the structure of possible microfossils detected by optical microscopy and to search for 'nanofossils'.
   Atomic Force Microscopy: to obtain 3D images of the structure, at nm-resolution, of possible microfossils and associated material.
   Macroscopic Observation: of stromatolite-type biosedimentary structures in freshly exposed scarps or in rock sections, using a telemicroscope or telescope with IR or Raman spectroscopic capability to confirm the mineral and carbonaceous constituents.
2. **Biogeochemical Indicators:** from the determination of the elemental abundances in reduced (organic) carbon remnants using a gas chromatograph/mass

spectrometer (GCMS) and alpha proton X-ray spectrometer (APXS) to give, for example, H/C ratio as index of aromaticity by assay of the 'organic' to 'carbonate' carbon content of fossil-containing sediments from an analysis of the volatile (hydrocarbon) component of the reduced carbon constituents in associated sedimentary material, using pyrolysis/gas chromatography/mass spectroscopy.

3. **Isotopic Indicators:** the enhancement of the ^{12}C isotope to ^{13}C during conversion of inorganic carbon to organic in autotrophic fixation is retained in biogenic carbon residues. It provides a distinct biomark when compared with the inorganic carbon pool. The pyrolyser/GCMS is used. An ion microprobe is used for microstructures. Similarly, D/H fractionation and $^{24}S/^{32}S$ isotope composition change between sulphides and sulphates can be indicators of earlier biological activity.
4. **Molecular Indicators:** certain distinct biochemical compounds, e.g. lipid type and pigments, may withstand degradation over very long periods as part of the kerogen residue. They may be extracted by organic solvents from residues and fossils for analysis (including chiral and isotopic) to obtain an indication of the nature of the original body.
5. **Chirality Indicator:** homochirality is characteristic of life and this structural feature is retained on death of the organism. Racemisation proceeds very slowly under cold dry conditions such as now exist on Mars. An optical rotation detector is used.
6. **Spectral Observations:** vibrational spectra of organic compounds provide a valuable analytical tool which can be applied to help unravel the nature of sedimentary carbonaceous samples of biological and abiotic origin. It also is able to identify the mineralogical content of the associated sediment. IR and Raman spectroscopy predominate and can be included in the microscopy.

Note: The references is for Appendix 5 may be found in the bibliographic entry for Chapter 5 in Appendix 1.

Appendix 6
The Rocket Equation

The famous Rocket Equation is expressed as:

$$\text{Final mass/launch mass} = \exp(\text{delta } V/g \times I_{sp})$$

Delta V is the change in vehicle velocity needed to get to the target and back, I_{sp} is the specific impulse which is a quantity directly associated with the type of propulsion technology used, such as chemical, electrical etc., and g is the constant of gravity at the Earth's surface. Figure D shows the return mass/launch mass ratio (vertical axis) as a function of required delta V for various targets (horizontal axis) with the specific impulse I_{sp} (associated with a given propulsion technology) as a variable parameter.

Various target examples are given, such as a round trip to geostationary orbit, slow round trip to Mars (400 days), fast round trip to Mars (180 days), Jupiter non-return (260 days) and Saturn non-return (540 days). From this it can be seen that

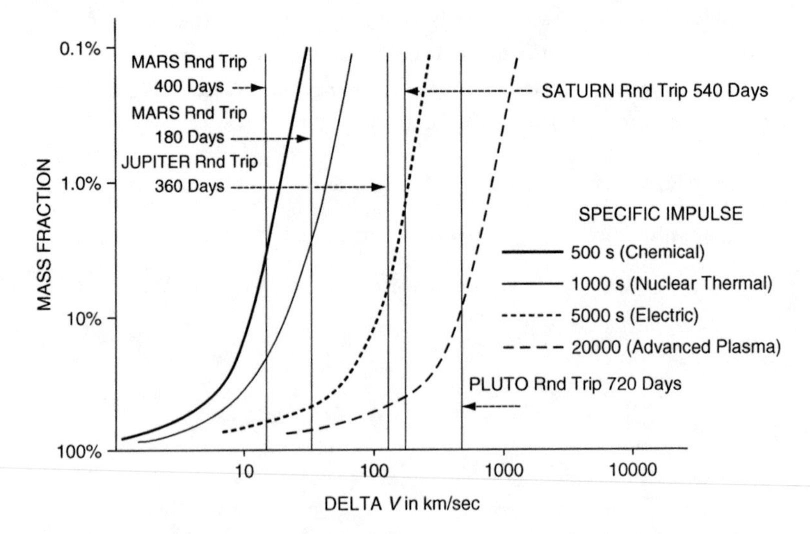

FIGURE C Return mars/launch mars ratio versus delta V for various targets with specific impulse I_{sp} as a parameter.

with chemical propulsion technology, a slow 400 day Mars return mission would allow a return mass ratio of about 5 %, whereas a fast 180-day mission would only allow 0.1% of launch mass to be returned. One can see from the exponential nature of the curves that chemical propulsion technology is pretty exhausted in terms of return mass at flight times much less than 400-days. The next technology, i.e. nuclear thermal, has an I_{sp} of about 1000 seconds when optimised and we can see that this could allow about 20 % recovery mass ratio for a 400-day Mars mission and 2 % for 180-day Mars mission. The next step up in terms of specific impulse is electric propulsion with an I_{sp} of about 5000 seconds. However, this does not tell the whole story since the thrust of a rocket motor is given by the product of the exhaust velocity and the mass flow rate:

thrust = exhaust velocity × mass flow rate

Because electric propulsion has to use an ionised gas as its propellant that is inherently of low density, the mass flow rate is low and hence electric propulsion motors have low thrust.

Appendix 7
Instrumentation

OPTICAL MICROSCOPY

Low magnification (20–80×)

Bulk documentation, colour analysis, texture, fossils, grain size analysis, grain shape analysis, selection of samples and/or sites for X-ray and microchemical analysis. A smooth surface on hard rocks is necessary.

High magnification (100–500×)

Details of major minerals, minor phases, accessories, surface of grains, twinning, mineral intergrowths, exsolution and micro reaction textures, microfossils. Ideally, hard rocks should have a polished surface.

At least two, preferably more, magnifications should be available on a well-centred revolver or sledge, or on two or more in-line microscopes. They could be equipped with colour diodes, a laser (Raman), IR source and UV light source.

ALPHA PROTON X-RAY SPECTROMETER (APXS)

The APXS can determine all elements except H and He with good accuracy. Hence, in the case of an ill-defined geometry, relative elemental abundances can easily be converted to absolute abundances. In particular, it seems to be able to analyse carbon down to about 0.1wt%. It can thus provide an important input on the puzzling question: where did the atmospheric CO_2 go to?

The APXS can determine the absolute concentration of oxygen, and it will also be able to shed light on the absolute oxidation state of the subsurface material as a function of depth.

MÖSSBAUER SPECTROMETER

Mössbauer spectroscopy is a powerful tool for analysing Fe-bearing minerals and iron compounds. In particular, the oxidation state of the iron and the Fe^{2+}/Fe^{3+} ratio in the soil and rocks will be determined directly. With Mössbauer spectroscopy, the Fe^{2+}/Fe^{3+} ratio can be determined for samples taken at different depths in the soil, which provides complementary information with respect to the H_2O_2 oxidant depth profile determination.

Biogenic magnetite very often (in most cases) shows up in Mössbauer spectra as a superparamagnetic component having strongly temperature-dependent Mössbauer parameters. Finding such signatures in the Mössbauer spectra will be an indication of previous biological activity. However, there are inorganic processes that also produce such nanophase magnetic particles, so care is needed in the interpretation of the data.

ION/ELECTRON PROBES, X-RAY SPECTROSCOPY

For chemical composition, the ion probe is heavy and an alternative could be electron probe analysis. Information on crystal structure could be provided by X-ray spectroscopy, if miniaturised for application to planetary exploration missions.

MOLECULAR SPECTROSCOPY

IR spectroscopy at 0.8–4.0 μm can determine the abundances of many types of minerals, including clays, hydrates, and carbonates. Carbonate materials such as magnesite, calcite and siderite can easily be distinguished from one another. All have near-IR features in the 1.7–2.5 μm range attributable to the CO_3- anion. Sulphates such as gypsum and anhydrite have absorptions in the 1.0–1.5 μm and 4.0–4.7 μm ranges. Ferrous absorptions of pyroxenes are also evident in the 1.0–2.0 μm region. The ratio of the 1 μm to 2 μm absorptions is diagnostic of the composition. Although olivine has been found in only one Martian meteorite, the structure of the broad absorption feature is also a diagnostic of composition. The absorptions of C–H and C–N bearing organics can be found throughout the 1.7–4.0 μm range. A spectral resolution of >100 $\lambda/\Delta\lambda$ would be sufficient. Good spatial resolution is needed to investigate variations within a sample and a resolution of 200 μ should be attainable.

A secondary objective for IR spectroscopy is investigation of the atmosphere. CO_2, H_2O and CO all have characteristic absorption bands in the IR. The variations of these gases and the dust cycle are strongly coupled in the Martian atmosphere and hence could provide significant additional information on the time variability of the atmosphere.

THERMAL IR SPECTROSCOPY

Thermal IR spectroscopy in the 6–9.5 μm region can be used to investigate specific absorptions/emissions of carbon-bearing molecules. Strong features are evident at

6.18 μm from the C–C and C–O bonds, 6.86 μm (-CH_2-, CH_2-scissor), 7.26 μm (CH bend) and 7.90 μm (O-H or C-H bend). This wavelength range could therefore provide a direct test for the presence of organics. Thermal emission spectroscopy can also determine whether geothermal or hydrothermal activity is occurring. Finding warm surface regions would be extremely important in the search for extant or extinct Martian micro-organisms. Is Mars geologically dead or is there sufficient heat produced to drive a subsurface hydrothermal circulation system? If this circulation is on a small scale, thermal emission spectroscopy may be a requirement in order to look in the optimum place for life.

Appendix 8
Terraforming Mars

FROM AUTONOMOUS BIOREGENERATIVE LIFE SUPPORT SYSTEMS TO '*ECOPOEISIS*' ON MARS

The establishment of a biological life support system on Mars is technically feasible and will probably be the conditio sine qua non for recycling the atmosphere and water in a Martian habitat and for producing food for the astronauts. This will include extended greenhouses on the surface of Mars, using as far as possible the natural resources of Mars itself.

In 1992, geneticist Bob Haynes and NASA geologist Chris McKay published a paper on the feasibility and motivation of implanting life on Mars (see Chapter 7 entries in Bibliography). They based their ideas on the results of the two Viking missions in 1976 which gave indications that the Martian atmosphere and surface might be modified to sustain life, even though the present Mars environment is extremely hostile. Taking the evolution of the Earth's biosphere as an example, they identified three main stages of biochemical and ecological innovation:

i. the chemical synthesis of the molecular components necessary for the formation of the first cellularised, membrane-associated replicating systems;
ii. the assembly of cells and the formation of anaerobic ecosystems involving the recycling of carbon; and
iii. the gradual accumulation of free oxygen in the atmosphere, the development of the contemporary biogeochemical cycles, and the diversification of the multicellular organisms.

All of these evolutionary steps occurred spontaneously on Earth. These three stages of biospheric evolution are described by the term '*ecopoeisis*', which implies the spontaneous formation of a biosphere on a planet. This word is taken from the Greek and means 'the making of an abode for life'. In contrast, the term 'terraforming' describes the directed fabrication of an Earth-like environment on another planet.

The scenario of '*ecopoeisis*' on Mars, developed by Haynes and McKay, includes three main phases: phase one would involve robotic and human exploration to determine whether sufficiently large and accessible volatiles inventories are available. The second phase would include planetary engineering designed to warm the planet, release liquid water and produce a thick carbon dioxide atmosphere. The time required to produce the energy needed for altering the climate of Mars is long. If an equivalent of 1 % of the incident solar energy could be used to warm the planet, it would require several hundred years to reach suitable life-supporting temperatures to a depth of a few meters in the Martian soil. A thick carbon dioxide atmosphere could already support some forms of life. However, the low concentration of nitrogen on Mars might be a problem for life as we know it. The third phase would include a programme of biological engineering designed to construct and implant pioneering microbial communities able to proliferate in the newly clement, although oxygen-less, Martian environment. However, this last step can only be performed if no indigenous Martian organisms are found or develop as liquid water becomes available. Although the scenario has several difficulties, it appears technologically feasible and it has been suggested by the authors that efforts to demonstrate the feasibility of '*ecopoeisis*' on Mars be one of the objectives of space research during the next century. Hence, the implantation of life on Mars is no longer confined to the realms of science fiction. However, before starting '*ecopoeisis*', all aspects of this enterprise including ethical ones need to be carefully weighed. These include the question whether we can ethically be allowed to drastically modify the climate of Mars, or whether we should preserve Mars in its present state for scientific exploration, such as is presently the case for Antarctica.

Appendix 9
Launch costs

A hard analysis of the situation regarding access to space eventually brings us to focus on one thing. The gateway, the portal, the philosophers' stone to opening the means to the planets and the stars is the reduction of launch costs to a fraction of their present values.

American engineer Peter A. Taylor (see Chapter 14 entry in Bibliography) has investigated this issue in 'Why are launch costs so high?' Here are some relevant facts derived from this analysis.

- It takes about 100 kg of kerosene and liquid oxygen to put 1 kg of payload in orbit. At about €0.40/kg this is only about €40/kg of payload to orbit.
- Lowest launch cost services on the commercial launch market are €10 000 – €20 000/kg to LEO.
  So it is not driven by propellant costs
- Aeroplanes cost to build about €300 – €600 per kg, so if we view expendable rockets as very expensive once-only-use aeroplanes and bear in mind that the aeroplanes weigh empty several times what their payload weighs, we get nearer the expendable rocket costs.
- One of the major drivers of cost is the impulse to avoid the *severity of the consequences of a failure*. A huge army of engineers is needed to ensure this. This is essentially driven by what is termed a lack of a 'continuous intact abort capability'. Basically, if an aeroplane has a technical problem it can usually land safely and have the problem solved. If a rocket launch vehicle has a problem it usually results in catastrophic failure – there is no capability to abort intact. Expendable vehicles have this problem by definition. Re-usables like the Space Shuttle have it partially, particularly in the first two minutes when the solid rocket-boosters are operating (Challenger disaster) or with damage to the thermal protection system (the infamous tiles) during launch (Columbia disaster), where even if NASA fully understood the problem – which it didn't – there was no intact abort capability. So huge armies of engineers are needed to minimise the risks of failure, driving the high costs. Taylor's conclusions are closed out by a statement that the best root cause is in the highly non-linear relationship between propellant mass fraction and payload capacity, i.e. the difficulty to build cheap expendables or safe re-usable vehicles. In this sense he believes that

> attempts to improve technology in the areas of decreasing dry mass and increasing specific impulse may have a cost reduction pay-off.

There is a real expectation that launch costs to low Earth orbit (LEO) will benefit from technological advances which will reduce the launch to LEO costs per kilogram by a factor of 10 within the next 20 years, thereby allowing the entry of commercial companies into transporting payloads into LEO and beyond at commercially attractive rates.

Index